Seed Testing

Seed Testing

Editor
Gurnam Singh

2014
Gene-Tech Books
New Delhi - 110 002

ISBN10 81-89729-89-6
ISBN13 978-81-89729-89-9

Published by : **GENE-TECH BOOKS**
4762-63/23, Ansari Road, Darya Ganj,
NEW DELHI - 110 002
Phone: 41562849
e-mail: genetechbooks@yahoo.co.in

Printed at : **Tarun Offset Printers**
Delhi

PRINTED IN INDIA

Preface

Seed is the basic and most important input for increasing agricultural productivity. The seed industry is developing at a faster rate with the added importance provided by the governments year after year for increasing the production of quality seeds. Seed testing plays a pivotal role in modern agriculture. The main objective of seed testing is to assess the actual planting value of the seed in terms of its germination capacity besides determination of percentage composition of the pure seeds, weed seeds, other crop species, extraneous matter and moisture content of the submitted sample.

Seed testing is the cornerstone of all other seed technologies. It is the means by which we measure the viability and all the physical factors that regulate the use and maintenance of seeds. Everything that is done with seeds should have some test information to guide the work and ensure high quality. Seed tests tell if a crop of seeds is worth collecting, if handling procedures are correct, and how many potential seedlings are available for regeneration.

This book seeks to facilitate the reader to grasp the basic principles of seed testing. Bringing forth a comprehensive analysis of the subject, it is hoped that this work fosters an enriching learning experience for the reader.

Editor

Contents

1

Seed Testing

Seed testing is the cornerstone of all other seed technologies. It is the means by which we measure the viability and all the physical factors that regulate the use and maintenance of seeds. Everything that is done with seeds should have some test information to guide the work and ensure high quality. Seed tests tell if a crop of seeds is worth collecting, if handling procedures are correct, and how many potential seedlings are available for regeneration.

The earliest form of seed analysis, the cut test, is still often used today. Before seeds are collected in the field, some seeds are cut open with a knife or razor blade to see if their internal tissues are fully developed and undamaged. This analysis is made more accurate in some cases by the use of a hand lens. It is also used for simple analysis during extraction and cleaning, or after germination to determine if the ungerminated seeds have deteriorated or remained dormant. Although the cut test is often very good at producing some information quickly, it is limited in the amount of information it can supply and it lacks accuracy compared to more sophisticated procedures. Therefore, it should never be taken as a substitute for a formal laboratory analysis.

Tests of the quality and other characteristics of seed need to be made at several stages during its progress from the parent tree to the seed bed. Several tests may be required at the seed processing depot, after extraction and cleaning and before the seeds are despatched to the nurseries or placed in storage. Testing of moisture content is essential for many species before they are stored for a long time, but is usually unnecessary if the seed is to be sown immediately in the nursery; on the other hand, partial drying may be needed before despatch of seed to a distant forest station, to reduce the risk of deterioration in transit, and any deliberate drying operation must be accompanied by the testing of moisture content.

Testing of germination or viability should be repeated at the end of the storage period if this is more than a few months and, in the case of long-term storage for conservation of genetic resources, testing should be done at intervals throughout the storage period. Efficiency and success in raising plants in the nursery and in their subsequent establishment in forest plantations depend to a great extent on the quality of the seeds used. It follows that foresters, nurserymen, seed dealers and others need accurate estimates of the quality of the seeds in which they deal or which form the basis of their afforestation projects. This is particularly important when seeds are bought and sold, or move between countries in international trade.

In practice, the number varies considerably from year to year and from seed lot to seed lot, so that it is necessary for the forester to have an accurate estimate of germination for every individual seed lot which he receives, if he is to fulfil planting targets while at the same time avoiding the waste of valuable seed caused by sowing too much. Similarly the nurseryman must have a good estimate of germination if he is to adjust sowing rates to produce the optimum spacing of seedlings in the seedbed.

Justice has defined the following objectives for developing rules for seed testing:

(a) to provide methods by which the quality of seed samples can be determined accurately;

(b) to prescribe methods by which seed analysts working in different laboratories in different countries throughout the world can obtain uniform results;

(c) to relate the laboratory results, insofar as possible, to planting value,

(d) to complete the tests within the shortest period of time possible, commensurate with the above-mentioned objectives; and

(e) to perform the tests in the most economical manner.

For the practising forester, the most important object of seed testing is the provision of an accurate estimate of the capacity of a given seedlot to produce healthy, vigorous plants suitable for field planting. The essence of good seed testing is the application of reliable standard methods of examination to ensure that uniform and reproducible results are obtained. Standardisation has been greatly facilitated through the adoption by a number of countries of the International Rules for Seed Testing formulated by the International Seed Testing Association (ISTA).

ISTA was founded in 1921, drew up its first set of Rules in 1931 and made major revisions of the Rules in 1953, 1966 and 1976. In the early stages ISTA's primary concern was for agricultural seeds, but trees and shrubs have gradually assumed greater importance and the 1976 Rules give (firm) "prescriptions" or (more tentative) "suggestions" as to testing methods appropriate for 61 different genera of these, compared with 26 in the 1953 Rules. The Rules are published in English, French and German.

Although tree species and genera from the tropics and southern hemisphere (notably eucalypts) have begun to figure in the ISTA lists, the balance is still overwhelmingly weighted towards north temperate species. Thus important species such as *Tectona grandis, Pinus patula, P. oocarpa* and *P. kesiya* are listed, but only in the list of "suggested" (non prescriptive) seed testing methods, while a recent survey has shown that 20 other important tropical species are not included at all; they include *Cupressus lusitanica, Gmelina arborea, Cordia alliodora* and the genera *Albizzia, Araucaria, Casuarina, Swietenia, Terminalia* and *Triplochiton*. Their omission simply reflects the lack of reliable information from research on the best methods of testing these species and genera.

ISTA testing methods involve the maintenance of controlled laboratory conditions and the use of some items of equipment which are relatively expensive. They are therefore well suited to large, well-equipped seed laboratories but are impracticable for small forest stations which have no laboratory and must carry out germination tests in the nursery or the office. *Lack of equipment is no reason to omit seed testing altogether.* For example simple germination tests carried out in the nursery can produce results which are perfectly satisfactory for local use. But precise details of method should always accompany results whenever this is not standardised, so that the seed user can interpret the results in relation to his own nursery conditions.

An up-to-date guide to the range of equipment available for seed testing is contained in the "Survey of Equipment and Supplies" published by ISTA. It lists equipment by type, use, model specification and supplier. The earlier directory, "Equipment and supplies for collecting, processing, storing and testing forest tree seed" includes seed testing equipment and gives the addresses of user laboratories as well as suppliers.

Sampling

When seed is tested, the sample on which the test is made has to be representative of the whole. No matter how accurate the technical work in the test, the results can only show the quality of the sample submitted for analysis. Therefore, every effort must be made to ensure that the submitted sample accurately reflects the composition of the whole seed lot. This is equally applicable, whether a whole series of tests is to be made in a laboratory, or a simple nursery test carried out to determine number of germinated seeds per kg of uncleaned seed.

Time spent testing carelessly drawn samples may be time wasted. Completely homogeneous seed lots would be easy to sample, but they do not exist. Common sense measures can be taken to reduce heterogeneity as much as possible, for example seed lots of the same species should not be mixed if they come from different origins (provenances), differ greatly in age or, in the case of the same introduced provenances, if they come from plantations growing on very different site types in the introducing country. But a seed lot may be heterogeneous even if collected from a homogeneous stand. Paul has given the simple example of pine seed transported over bad roads in the back of a landrover.

The bumps and vibrations to which it is subjected will cause a settling out of different types of material. All the small and empty seeds and other light rubbish will eventually reach the surface layers and a sample taken from the top alone will give a completely false impression of the potenual performance of the seeds as a whole.

Mixing

If the seed lot is a small one consisting of only a few kg of seed, it is possible to improve its homogeneity by thorough mixing of the whole lot before taking a sample. In the case of very large seed lots transported and stored in many different containers, mixing of the whole seed lot is impracticable. The methods of mixing which are quoted below from Paul, are equally applicable to small entire seed lots or to composite samples. "*Methods of mixing*. There are two simple ways in which seed may be thoroughly mixed.

i) *Mixing with the aid of a mechanical divider.* Mechanical dividers are used to reduce the size of seed lots or samples by repetitive halving. Their operation is described below. The divider has the added advantage that it can also be used to mix seed in the following manner:

   1. Pass entire seed lot through divider.
   2. Take the two equal portions and feed them simultaneously into the divider.
   3. Repeat operation No. 2.
   4. Repeat operation No. 2 once more.
   5. Take the two equal portions and pour them simultaneously into the storage container.

ii) *Handmixing*. It is surprising how difficult it can be to achieve a homogeneous seed lot using this method. The entire seed lot must be spread out on a sheet of paper or some other suitable smooth surface and mixed by scooping the seed from side to side and from top to bottom. After thoroughly mixing, spread the seed out evenly and divide the lot into four equal parts. Place these separately in four containers and with the aid of an assistant pour them simultaneously into the storage container. Repeat the spreading, quartering and pouring process another two times."

Seed Triers

For a large seed lot contained in a number of different containers, seed triers are used to take small sub samples from different parts of the lot. All of these "primary" samples are mixed together to form a "composite" or bulk sample, which is later reduced in size by successive dividings until it is small enough to form the "working" sample on which the various tests are carried out.

A seed trier is a probe, long enough to reach all areas of the seed bag and designed to remove an equal volume of seed from each area through which it travels. Most commonly used is a sleeve-type trier which consists of a hollow brass tube inside a closely fitting outer sleeve. The tube and sleeve have open slots in their walls so that, when the tube is turned or slid, the slots in the tube and the sleeve coincide and seeds can flow into the tube; when the tube is given a half turn or slid in the opposite direction, the openings are closed.

Seed triers should have a series of separate compartments and not just a single long tube with several openings. Tubes are available in varying lengths and diameters to cater for different sizes of container and kinds of seeds. Triers known as "thief" triers, with a single open slot, should be avoided, as they can damage the seeds. Ideally, the primary samples should be distributed proportionally to the volumes of the different parts of the seed lot. For example, if a lot is contained in ten 10-kg containers and ten 20-kg containers, the 20-kg containers should contribute two thirds of the composite sample and the 10-kg containers only one third.

For seed lots in containers of uniform size, ISTA has detailed prescriptions as to the number of primary samples appropriate for varying numbers of containers, e.g. for 6 – 30 containers at least one in three containers is to be sampled and never less than five. Large seeds and others that are not free-flowing are not suitable for sampling by seed trier. They should be sampled by thrusting the hand into the seeds and removing small portions. The hand should be inserted flat with the fingers extended together. The fingers should be kept together as the hand is closed and withdrawn. It is difficult to sample with this method deeper than about 40 cm and containers may have to be partially emptied to facilitate sampling.

Usually the composite or bulk sample for testing needs to be reduced to a working sample of standard weight. Whenever possible the sample should be subdivided by a mechanical divider to reduce any operator bias in the dividing. The non-mechanical halving method is described here for use if a mechanical divider is not available, but it is the least desirable method for most types of seed. Manual methods must, however, be used for seeds which are not free-flowing.

Non-mechanical Methods of Dividing

Halving method. The sample is placed on a clean surface and thoroughly mixed by hand; it is then divided into quarters with a sharp-edged spatula and opposite quarters discarded. The process is repeated until a final sample of the approximate weight required is obtained.

Random cup method. A series of small cups or thimbles are arranged on a tray in a definite pattern and the sample is poured systematically over this area. The working sample is obtained from randomly selected containers. A modification of this method uses a tray divided into an equal number of square compartments, every alternate one of which has no bottom.

Mechanical Methods of Dividing

Most mechanical dividers are designed to split the sample into two approximately equal parts. The working sample is obtained by repeatedly dividing the bulk sample until it reaches the required weight. ISTA recommends three types of dividers as being suitable equipment. They are the conical divider, the soil divider, and the centrifugal divider (Gamet type). They are briefly described as follows:

- *Boerner sampler:* This divider is made in different sizes. The essential parts consist of a hopper, inverted cone, and a series of baffles directing the seeds into two spouts. The baffles form alternate channels and spaces of equal width. They are arranged in a circle at their summit and are directed inward and downward, the channels leading to one spout and the spaces to an opposite spout. A valve or gate at the base of the hopper controls the seed flow. When the valve is opened the seeds fall by gravity over the inverted cone where they are evenly distributed to the channels and spaces; then, they pass through the spouts into the seed pans below.

In Zimbabwe a similar arrangement on a larger scale is used for dividing and mixing large seed lots. From the inverted cone there are 8 outlets which pass the seed into buckets, and a battery of 8 buckets is available for each outlet. The buckets hold 6 kg of pine seed each, so that a seed lot of up to 384 kg can be handled in any one dividing and remixing operation.

– *Soil divider*: A simple divider, built on the same principle as the conical divider, is the so-called soil divider or riffle divider. The channels are here arranged in a straight row instead of a circle as in the conical divider. The soil divider consists of a hopper with attached channels or ducts, a frame to hold the hopper, two receiving pans and a pouring pan. In using the divider the seed is scattered fairly evenly in a pouring pan and poured in at approximately equal rates along the entire length of the hopper. The divider is suitable for large-seeded and chaffy species, but suitable types for small-seeded species can also be made.

– *Gamet divider*: The Gamet divider makes use of centrifugal force to mix and scatter the seeds over the dividing surface. The seeds flow downward from a hopper onto a shallow rubber cup below. Upon rotation of the rubber cup by an electric motor the seeds are thrown out by centrifugal force and fall downward. The circle or area where the seeds fall is equally divided into two parts by a sharp-edged, stationary baffle so that one-half of the seeds fall in one spout and the other half in the other spout. In using this divider care must be exercised in dividing very small samples as it is possible that a majority of seeds may be thrown out in one spout. Hardin et al. compared the effectiveness of the Boerner and Gamet dividers and concluded that *if used correctly*, the Gamet divider yields a slightly more accurate subsample.

Sample Weight

The weight of the working sample depends on the seed size of the species in question. The ISTA rules are intended to provide a minimum of 2500 seeds for all except the very large-seeded species, for which a minimum of 500 seeds is prescribed. These quantities are considered sufficient for most of the usual tests (purity, authenticity, seed weight, germination or viability), but an additional 10 g of most species is needed if moisture content is to be determined.

Among tropical species listed by ISTA the range of prescribed minimum size of working sample is from 2 g for *Eucalyptus deglupta*, which averages 4,000 viable seeds per gm of "seed plus chaff" to 1 kg for *Tectona grandis* which averages 2,000 fruits per kg. When seed is despatched to an independent seed laboratory for testing, ISTA recommends that the weight of the submitted sample be double that of the working sample.

Purity Testing

Tree seed samples can contain impurities such as weed seeds, seeds of other tree species, detached seed structures, leaf particles and other material. The object of purity analysis is to determine the composition by weight of the sample being tested. To do this the sample is separated into component parts. When purity analysis is done, it is the first test to be carried out because subsequent tests are made only on the pure seed component. *Pure seed* refers to the species under consideration and in addition to mature, undamaged seed includes: undersized shrivelled, immature and germinated seeds, provided they can be definitely identified as the species under consideration; and pieces resulting from breakage that are more than one-half their original size.

Seeds of *Leguminosae* and *Coniferae* with the seed coats entirely removed are regarded as inert matter. Other components of the sample may include *Other seeds* of all species except that under test and *Inert matter.* Inert matter consists of pieces of broken or damaged seeds less than half the original size, wings of coniferous species, leguminous and coniferous seeds with the seedcoat entirely removed and *other matter* such as fragments of leaves, twigs, stones, soil. In Coniferae (except *Chamaecyparis, Cupressus* and *Thuja*) any remaining seed wings not already removed in cleaning should be detached and classed as inert matter.

The working sample containing all the impurities is weighed and then the pure seed is removed and weighed separately. ISTA rules prescribe that weighing be done in gms to the minimum number of decimal places necessary to calculate the percentage of its component parts to one decimal place. The percentage of pure seed is calculated as follows:

$$\text{Purity } \% = \frac{\text{Weight of "pure" seed}}{\text{Total weight of original sample}} \times 100$$

When a laboratory is testing a large number of samples for purity, the purity analyst usually examines and separates seed samples on a "workboard" placed on top of a desk or table. The workboard can be adjusted to a desired height, usually 7 - 15 cm above table height. The analyst's equipment should permit him to do his work with a minimum of effort and time, and also with a minimum of eye strain.

Equipment required for making purity analysis is as follows: forceps and spatula for handling, separating, and pushing seeds over a surface; wide-field hand lens of 5X to 7X magnification; reading magnifiers relatively free of curvature and distortion, wide-field stereoscopic microscope with a range of magnifications from 10X to 75X; scale with capacity up to 1000 g and sensitive to 0.5 g; torsion balance with capacity of 120 g and accurate to 0.01 g ; rapid-acting chemical balance accurate to 1 mg; small seed containers to hold separations; and seed blower. An assortment of small sieves can help in eliminating many impurities.

Analyses of many samples can be facilitated by removing the lightweight material with the aid of a seed blower. Other methods reported by Turnbull as reducing hand labour are specific gravity separation by differences in density, separations by electrostatic charges, a vibrating table working on the basis of seed weight, surface texture and shape, and an X-ray radiograph technique.

The pure seed component from the purity analysis can be subsampled for the germination test, as well as for the determination of seed weight. Since the germination test is based on pure seed, it can be readily seen that the purity analysis and germination test are complementary to each other. The production potential of a seed lot can be determined only when the purity analysis and the germination tests are considered together.

Seed Weight

Measurement of seed weight is made on the pure seed

component separated by the purity test. Weight is normally expressed as the weight of 1000 pure seeds. This figure can be readily converted to number of pure seeds per g or per kg as required. Weight may be determined simply by counting out 1000 seeds and weighing them, but the use of several smaller samples enables the analyst to estimate variation within the sample.

ISTA prescribes 8 replicates of 100 seeds each, from which the standard deviation and coefficient of variation may be calculated, as well as the mean. If the co-efficient of variation is less than 4, the mean is accepted, but, if it is more, 8 additional replicates are prescribed, a new standard deviation for the 16 replicates calculated and any replicate that diverges from the mean by more than twice the standard deviation is discarded before calculating the final mean for the sample. The counting of seeds for weight may be done by hand or with the assistance of counting boards or vacuum or electronic counter. Some essential features of seed counting devices are briefly described below.

Counting boards are useful for relatively large seeds of a regular shape, such as *Pinus pinea* or some leguminous species. The counter consists of two boards, a stationary top board with perforations and a thin solid board below, which serves as a false bottom. In operating the counter, a quantity of seed is placed at the corner, the board tilted and moved so that one seed goes into each perforation. The board can then be emptied and the known number of seeds from it weighed.

Alternatively, if the seeds are being counted in preparation for a germination test, the counter can be placed on the substrate and the bottom is withdrawn permitting the seed to fall through at a uniform spacing. In some types there is a second set of perforations in the bottom board. By slipping the top section about one centimetre, the seed falls through both boards onto the germination bed. In other types, the counting apparatus is perforated by holes too small to allow the seeds to pass through but suitable to allow a flow of air to suck the seeds into position. In connection with tests on soil trays, a simple counting perforated board can also be employed. The board is first placed over the tray. The seeds are placed in the perforations and the board is lifted leaving the seeds on the tray.

Vacuum counters can be used with seeds of various sizes and shapes. For very small and irregularly shaped seeds some difficulties may occur. They are not suitable for very large seeds, such as many nuts. A vacuum counting device has three essential parts: a vacuum system including pipes; counting heads; a quick release valve. A small house vacuum cleaner with a home-made head may be used, but special power units are generally preferable (1/4 to 1/2 hp electric motors). Perforated heads, with evenly spaced holes (generally 100) to house the seed, hold the seed by suction. When suction is switched off, the seeds are released for weighing or onto the substrate for germination testing.

Counting plates or heads are of a square, rectangular or circular form, according to the shape of the substrate employed. Espacement and diameter of holes should be proportionate to the seed size. Heads can be of plastic, chromium, brass or aluminium. The face plate should be relatively opaque and of a colour which contrasts with the seeds being counted, thus making it easier to check that all holes are properly filled. A good valve is needed to allow a quick release of seed, as well as to avoid sucking up too large a proportion of the lightest seeds.

The risk of picking up an excessive proportion of light, empty seeds in the counted sample may be reduced by inverting the head so that the holes are upwards and then pouring a large sample of seeds over it. Operation of vacuum counters with plastic heads may be adversely affected by static electricity. In the modern types of electronic seed counters a vibrating feed bowl moves seeds past an electric eye in single file. The machine can be programmed to count a certain sample size, then cut itself off.

Reservations have been expressed about the accuracy of vacuum and electronic counters on the grounds that they select a biassed sample. Manual counting boards, combined with careful random sampling, are likely to be more accurate, although considerably slower. The 1000 pure seed weight can be converted to seeds per g or per kg as follows:

$$\text{Number of seeds per g} = \frac{1000}{\text{Weighting of 1000 seeds}}$$

or

$$\text{Number of seeds per kg} = \frac{1000 \times 1000}{\text{Weighting of 1000 seeds}}$$

If the sample counted is other than 1000 seeds, the appropriate formula is:

$$\text{Number of seeds per g} = \frac{\text{No. of seeds in sample}}{\text{Weight of sampling}}$$

or

$$\text{Number of seeds per kg} = \frac{\text{No. of seeds in sample} \times 1000}{\text{Weight of sampling}}$$

Within a given species full seeds have a higher specific gravity and a higher germination rate than seeds of the same size but empty or partially filled. Large seeds weigh more per seed than small ones of the same specific gravity and, because they contain larger food reserves, they are likely to germinate better and produce initially more vigorous seedlings. Goor and Barney have reported that large seeds of *Eucalyptus citriodora* had a higher germination rate than medium-sized seeds which in turn had a higher rate than small seeds.

Number of pure seeds per unit weight is therefore not in itself a good guide to potential for plant production and must be complemented by germination tests or indirect tests of viability. The effect of seed size on the growth of germinated seedlings of eucalypts usually persists for 8–14 weeks after sowing, thereafter other factors may become more important.

Seed Germination

Of all the quality measurements of seed lots, none is more important than the potential germination of the seeds. The main aim of a laboratory germination test is to estimate the *maximum* number of seeds which can germinate in *optimum* conditions. The use of standardised ideal conditions in the laboratory such as those prescribed by ISTA ensures that results obtained from a given seed lot in one laboratory should be identical with those obtained from any other laboratory in the same or in other countries.

A common standard for assessing germinative potential is very valuable for seed moving in international trade. On the other hand, it is clear that results obtained under ideal controlled conditions in the laboratory are not directly applicable in the field nursery, where only limited control over environmental conditions is possible. Each nurseryman should apply his own correction factor, derived from experience over the years, to convert the germination potential of a seed lot, as determined from laboratory tests, to the actual field germination he can expect under his own local conditions.

At the other extreme, a nurseryman may prefer to carry out his own germination test in his own nursery in advance of large scale sowing. The results of the test should be directly applicable to the later sowings of the same seed lot in the same nursery, but would not be applicable to other nurseries. However, there is sometimes insufficient time for germination tests in advance of the main sowing and managers of large operational nurseries may be reluctant to involve themselves in small scale research.

Between the two extremes lies the case of the small silvicultural research station which lacks the laboratory facilities to comply with ISTA prescriptions for testing, but which carries out germination tests in the nursery on bulk seed supplies before distributing them to afforestation projects throughout the country. Here again the local forester or nurseryman must apply his own correction factor to convert actual germination figures from the research nursery to expected germination figures in his large operational nursery.

Germination is defined as the emergence and development from the seed embryo of those essential structures which are indicative of the seed's capacity to produce a normal plant under favourable conditions. Germination is expressed as the percentage of pure seeds which produces normal seedlings or as the number of seeds germinating per unit weight of the sample.

In the laboratory the environmental conditions, including moisture, temperature, aeration and light, must not only be specific enough to initiate germination but also favourable for the development of the seedlings to a stage where interpretation of normal and abnormal types may be made. With a few exceptions, all germination tests should be made with pure seeds separated by the purity test. The pure seed must be well mixed and counted at random into replicates. The seeds should then be spaced uniformly on the test substrate.

Normally one test consists of 400 seeds in 4 replicates of 100 seeds each, but if 100 seeds overcrowd the test substrate, the replicates may be broken down into a larger number of smaller replicates of 50 or 25 seeds each. A general recommendation is to leave 1.5 to 5 times the normal seed width or diameter between seeds, in order to discourage the spread of fungal moulds.

The exceptions are very small-seeded species in which it is impossible (some species of *Eucalyptus*) or very difficult (*Alnus, Betula, Populus, Salix*) to separate the seeds from the accompanying inert matter or chaff. In these cases the test is made with the same number of replicates but the replicates are of equal weight rather than an equal number of seeds.

Germination Equipment

The kind of germinator can be selected according to the type and amount of seeds being tested and it is acceptable providing it can give good control of prescribed temperature, moisture, and light conditions. Germinators range in size from small or portable individual units through cabinets of various sizes and large germination tables of the Jacobsen Copenhagen type, to walk-in germination rooms. The major types recommended by ISTA are as follows:

Jacobsen and Rodewald apparatus: In Europe, a germinator called the *Jacobsen apparatus* or *Copenhagen tank* is commonly used. It consists of a bath containing water, the temperature of which may be controlled by thermostats. The seeds are distributed on top of paper and placed on metal or glass strips suspended about 5–7 cm above the bath; paper or cotton wicks lie beneath the substrate and pass through slots into the water below.

The moisture content of the substrate may be adjusted by altering the water level, and so increasing or decreasing the distance between the seed bed and the water. High humidity is maintained around the seed either by covering the entire tank with a transparent lid or placing an inverted plastic funnel with a hole in the conical end over each germination pad. The apparatus may be exposed to natural daylight but artificial light is usually desirable. One of the disadvantages of the standard Copenhagen tank is the absence of direct temperature control at the seed bed. In the tropics it tends to overheat.

Improved models have been developed, such as the one by Overaa which uses hollow stainless-steel strips, through which the water for heating and cooling is circulated. Other disadvantages are that the equipment requires a large amount of floor space for the number of tests accommodated. Also, the handling of covers and substrata appears to be more cumbersome than moving a lightweight tray with paper substrate.

The Rodewald apparatus consists of a zinc box covered with glass in which the seeds are exposed to direct or diffuse light. The bottom of the apparatus contains water over which a layer of moist sand is placed on a tray. The seed bed consists of unglazed porcelain dishes placed in the moist sand, or the dishes may be placed directly in the water. The temperature of the water is controlled thermostatically and the sand is moistened through wicks reaching into the water bath. A sand substrate is not desirable for species requiring alternating temperatures because of its slow adjustment to a change in temperature.

Germination cabinet: Another common type of apparatus is the closed chamber for germinating seeds in darkness, diffused light, or direct light. This type of germinator often consists of a double walled cabinet, suitably insulated against temperature changes by an air jacket or layer of insulating material. It is equipped with appropriate tray slides to accomodate the type of germination trays preferred by individual laboratories. The modern germination chamber has provision for both heating and cooling.

Usually, water is chilled and passed between the walls of the chamber or through cooling pipes placed along the inner walls of the chamber. Either the air or a reservoir of water in the bottom of the chamber may be heated. In such cabinets the temperature can be regulated at any desired setting between approximately 8°C and 40°C. Inexpensive germination cabinets are sometimes constructed locally. A germination cabinet should meet as far as possible the following requirements:

- *air humidity* - as high as possible but preferably never lower than 90%, to prevent the seed substrate from drying up;
- *air temperature* - adjustable between 10°C and 35°C, the temperature all over the working section of the cabinet over a number of days to be uniform in each of the temperature regimes, within plus or minus 1°C;

- *light* - uniform illumination of the trays, intensity 750 to 1250 lux at the level of the seed;
- *air movement* - as low as possible to prevent the seeds from drying up;
- *fresh air supply* - small, of the order of one air change per hour, which should be sufficient to remove the carbon dioxide produced by the germinating seeds;
- *day and night cycling* - a rapid initial temperature drop to be achieved in 30 minutes during the day to night change, reaching the ultimate night temperature in one hour, a similar requirement to hold for the night to day change;
- *condensation* - not allowed.

However, when cabinets operate with alternating temperatures it is very difficult to maintain the high humidity and the substrates dry out; this necessitates frequent watering, which adversely affects the germination results and adds to the manhours spent on the test.

Room germinator. When large numbers of tests are made, whole rooms may be used as germinators, with temperature, humidity and light controlled. The room germinator is a modification of the cabinet. It is constructed on the same principle as the cabinet, but is large enough to permit workers to enter it and place the tests along either side of a central passageway. Usually, fans are required to prevent stratification of temperature and special equipment is necessary to maintain a high relative humidity.

Another modification is the combination room-cabinet germinator. The temperature of the entire room is kept as low as the lowest temperature required. Germination cabinets are placed in this room and heated, individually, by electricity to maintain the various temperatures required. Either constant or alternating temperatures may be obtained by this type of germinator.

Portable germination boxes. A simple and versatile germinator consists of a number of transparent plastic boxes with lids, which can be stacked one upon another.

Selection of a germinator. Most forest seed laboratories with a moderate through-put of samples use either the cabinet-type germinators or germination tables. The advantages of using cabinets instead of germination tables of the Copenhagen type are a saving in space and, in well constructed cabinets, the more precise control of temperature and humidity. It should be noted that a single-level Copenhagen tank requires five to eight times the floor space occupied by a cabinet of equal capacity. Where research is being carried out on germination conditions, or where testing covers a wide variety of seeds, the flexibility provided by several cabinets operating at different temperatures and/or light conditions can be very advantageous.

For nurseries or small research institutes concerned with carrying out simple tests of germination without maintaining strict control of temperature and light, the flexibility, storability and cheapness of plastic germination boxes have much to recommend them. They can also be used in combination with incubators in which light and temperature are controlled. The optimum conditions for different stages of germination and seedling growth are not identical and may even vary for different seeds within the same seed lot. The aim of much seed testing research has therefore been to determine a combination of conditions which will give the most regular, rapid and complete germination for the majority of the same species.

Substrate: Soil is rarely used as a substrate for germination tests because each sample will vary greatly in physical, chemical and biological properties. While the germination results might be more comparable with field conditions, the lack of reproducibility and difficulty in comparing tests of different seed lots render it unsuitable. Artificial media are much more easily standardised. Most laboratory tests on small seeded species are made on paper. Other materials used include sand, granulated peat moss and expanded mica (Vermiculite and Terralite). The main requirements for the substrate are:

- non-toxic to the germinating seedlings
- free of fungi and other micro-organisms
- of porous texture to enable adequate aeration and moisture for the germinating seeds.

The choice of media on which the seeds are placed for germination depends on the equipment, the species, the working conditions and the experience of the operator. It is prescribed in the ISTA Rules for a number of forest trees. Filter paper, heavy box paper, or any other absorbent paper may be linked with a filter paper or cotton wick dipped into water, or placed on top of a layer of sand or vermiculite.

Cellulose paper is becoming more commonly used as a germination medium because it is easier to handle than sand, but still permits penetration of the radicle, thus providing for better evaluation of abnormal germination. Also moisture content does not tend to become stratified within the paper medium as it does with sand. The best paper substrates are germination blotters, paper towelling, laboratory filter paper and creped cellulose paper wadding. Any paper substrates should be checked for presence of toxic chemicals.

The 1976 ISTA rules give detailed specifications for paper and towels, including weight, bursting strength, capillary rise and acidity. Seeds without a specific light requirement may be placed on top of paper or between folded paper. Folded paper increases the surface contact area between the seeds and the moisture supply. Large seeds may be rolled in paper towels which are then placed in an upright position; this allows the roots to grow downwards and avoids tangling of the roots. Rolled or folded paper packs may be kept moist without the necessity of a wick dipping into water.

Rolled seeds may be stored on glass plates above, but not in contact with, a water bath and covered by a polythene sheet. Any rolls which show signs of drying out may be sprinkled with water. Use of rolled paper is a very rapid and convenient method but may cause curling of the radicles. This is of no concern if the germinated seeds are to be discarded after testing. For operational germination of large quantities of seed, however, the slightly slower folded paper method is preferable. This is used with success as standard practice for *Pinus* and *Eucalyptus* spp. in Thailand.

Sand is not suitable for very small seeds because they are difficult to locate, but it is widely used for larger seeds. Sand can be sterilised and fungi develop on it less freely than on paper. It also provides a good contact between the seed and moisture because the seeds can be pressed into the medium. A commonsense rule is to cover the seed with a thickness of sand not less than the length of the seed on its longest axis.

ISTA recommends a thickness of 1 - 2 cm according to seed size and prescribes that sand particle sizes should be between 0.05 and 0.8 mm and pH between 6.0 and 7.5. Sand as a germination substrate is preferred for tree species that have a longer germinative period, for instance, *Rosa* spp., *Pinus caribaea*, *P. elliottii*, and *P. palustris*, or with larger seeds, for example, *P. pinea* and *Quercus* spp. A sand-perlite medium permits leaching of the water which can be important when testing repellent-coated seed, but may be a disadvantage when testing seeds which are sensitive to drying.

Moisture and Aeration: The moisture level of the substrate has been suggested as one of the major causes of variation in seed research results. The ISTA Rules specify that a sand substrate be moistened depending on its characteristics and the size of the seed to be tested and suggest that a moisture level of 50 - 60% of the water-holding capacity of the sand is suitable for several groups of agricultural seeds. A paper substrate must not be so wet that a film of water forms around the seed. As a generalisation, the substrate must be moist enough at all times to supply the necessary moisture, but excessive moisture will restrict aeration. All tests must be examined daily to ensure that the moisture content of the substrate is near optimum. The water should be reasonably free of impurities.

Temperature control. In laboratory seed testing the important temperature is that at the level of the seed. The temperature will vary for different species and appropriate values are listed for a large number of tree species in the International Rules for Seed Testing.

Temperature is one of the most critical factors in the laboratory germination of seeds and must be regularly checked. When alternating temperatures are required the test is usually held at the low temperature for 16 hours and at the high temperature for 8 hours each day.

Temperature of 20° and 30°C alternating are commonly prescribed for tree species. Although natural fluctuations between day and night temperatures are less in lowland moist tropical forests than in other forest types, alternating temperatures may still affect germination of tropical species. In one experiment of *Terminalia ivorensis* in Nigeria, alternating temperatures of 34° C and 24° C gave 93% germination in 41 days compared with 27% at a constant temperature of 30° C, both with continuous light.

Light. Light is required for germination in many tree seeds. Fluorescent light is as effective as natural daylight and is preferred in testing, because the wavelength and intensity can be standardised, within limits. Cool white fluorescent lamps are recommended because of their light quality and low heat emission: Light should be evenly distributed over the tests, with an intensity within the range 750 - 1250 lux. Seeds should be subjected to light for only part of the test period, 8 hours in every 24 hours is usual but a longer or shorter period may be beneficial to seeds of some species.

Cotrolling fungi in germination tests. Laboratory practices that will minimise the spread of fungi include proper spacing of seeds, control of temperature, removing decayed seeds, proper aeration, and keeping the substrate only just wet enough to permit germination. Sterilisation of laboratory equipment and the periodic disinfection of germination cabinets and other apparatus will assist. As a general rule, seed is not disinfected because seeds which decay are usually of poor quality.

A number of experiments carried out with different methods of disinfecting the seed just before or during germination have shown no reliable improvement in germination percentage. A small quantity of fungicide added to the water used to moisten paper rolls has, however, been found beneficial in testing conifer seeds in Denmark. In Australia spraying of Karathane fungicide at a concentration of 0.8 g in one litre of distilled water has also proved effective when testing eucalypt seed on moist filter paper over vermiculite.

A seed is considered to have germinated after the emergence and development from the seed embryo of those essential structures which are indicative of the seed's capacity to produce a normal seedling under favourable conditions. Abnormal seedlings are not included in the germination count because they rarely survive to produce plants. The ISTA Rules recognise four groups of abnormal seedlings:

(a) damaged seedlings,

(b) deformed seedlings,

(c) decayed seedlings,

(d) seedlings with unusual hypocotyl development.

These groups and their characteristics are defined in detail in the ISTA Rules. In a laboratory test the majority of the normal seedlings are usually removed at the interim counts, but the assessment of many of the doubtful and abnormal seedlings must be left until the end of the test, to ensure that slower growing but otherwise normal seedlings are not incorrectly classified. For many species the initial count is carried out one week after starting the test and assessments may be carried out at weekly intervals until the test is ended. More frequent assessment is needed if a more precise picture is needed of the speed of germination.

At the end of the test period, all remaining ungerminated seeds should be cut and examined, and the number of fresh, firm and possibly viable seeds recorded.

At the same time observations may be made on the condition of unsound ungerminated seeds e.g. an abnormally high incidence of insect–damaged or mechanically damaged seeds which would indicate the need for improvements in seed hygiene or seed processing methods. The record of the germination test commonly shows the percentage of germinated and the percentage of ungerminated but apparently sound seeds separately, e.g.

Germination %	= 82%
Sound ungerminated seeds	= 6%
Viability %	= 82% + 6% (= 88%)

Many species require no pretreatment and some types of seedcoat dormancy can be successfully treated in a matter of hours. But the prescribed pretreatment of *Tectona* takes 18 days and some prescribed prechilling treatments 3 - 9 months. When planning a testing programme it is necessary to take account of pretreatment as well as testing time. In extreme cases it may be necessary to replace the germination test by one of the indirect tests of viability described below.

Germination Energy

Germination energy has been defined in more than one way: (1) The percent, by number, of seeds in a given sample which germinate within a given period (defined as the energy period) e.g. in 7 or 14 days, under optimum or stated conditions or (2) The percent, by number, of seeds in a given sample which germinate up to the time of peak germination, generally taken as the highest number of germinations in a 24 hour period. In both definitions the length of the energy period is considerably shorter than the full test period prescribed by ISTA.

Germinative energy is a measure of the speed of germination and hence, it is assumed, of the vigour of the seed and of the seedling which it produces. The interest in germinative energy is based on a theory that only those seeds which germinate rapidly and vigorously under the favourable conditions of the laboratory are likely to be capable of producing vigorous seedlings in field conditions, where weak or delayed germination is often fatal.

Another method of comparing germination energy of different seed lots is to record "Rate of Germination" i.e. the number of days required to attain 50% of germination capacity. The shorter the period, the greater the germination energy. Yet another method is to evaluate the stage of development of germinated seeds and classify them into a number of development or vigour classes. For example Wang used seven classes for germinated normal seeds of *Picea glauca,* in addition to ungerminated and abnormally germinated seeds; they ranged from seedlings with healthy root, fully developed hypocotyl and seedcoat completely shed to seeds with seedcoat burst open but as yet no radicle emergence.

Germination Value

The concept of Germination Value aims to combine in a single figure an expression of total germination at the end of the test period with an expression of germination energy or speed of germination. Total germination is expressed as (final) *Mean Daily Germination* (MDG), calculated as the cumulative percentage of full seed germination at the end of the test, divided by the number of days from sowing to the end of the test. Speed of germination is expressed as *Peak Value*, which is the maximum mean daily germination (cumulative percentage of full seed germination divided by number of days elapsed since sowing date) reached at any time during the period of the test. *Germination Value* (GV) can then be calculated from the formula.

$$GV = \text{(final) } MDG \times PV$$

Germination Value, as an integrated measure of seed quality, has been used by several tropical seed workers e.g. for *Terminalia ivorensis* and for *Pinus kesiya*. An alternative method of calculating Germination Value has been proposed by Djavanshir and Pourbeik, who found that it was more closely related to survival of plants in field nurseries than was Czabator's method, in the case of *Pinus ponderosa* and *P. eldarica* in Iran. The formula they proposed was:

$$GV = (\Sigma DGS / N) \times \frac{GP}{10}$$

where:

GV = Germination value

GP = Germination percent at the end of the test

DGS = Daily germination speed, obtained by dividing the cumulative germination percent by the number of days since sowing

ΣDGS = The total obtained by adding every DGS figure obtained from the daily counts

N = The number of daily counts, starting from the date of first germination.

The calculations required are somewhat lengthier than those of the Czabator method and, for many patterns of germination, the simpler method is likely to give sufficiently accurate comparisons between seed lots.

Germination Testing in the Nursery

A good example of a suitable method for testing germination in the nursery is the procedure recommended for tropical pines in West Malaysia.

Take 400 pure seed and divide into 4 samples each of 100 seed. Take 4 wooden or plastic boxes (opaque flexible plastic sandwich boxes are very good) of dimensions 30 × 30 cm and of depth about 10 cm and fill them slightly overfull with sieved sand to the top of the sides. Wet the sand by applying 0.5 litres of water to each box. Level the sand across the top of each box and make drills at 2.5 cm intervals and of depth 6 mm. Sow the seed (100) at intervals of 2.5 cm and gently press them on to the sand in the drill. Cover to a depth of 6 mm (3 mm for *P. oocarpa*) with sand which has been passed through a Mesh No. 12 (12/64" or 4.76 mm) but retained on a Mesh No. 8 (Size 8/64" or 3.18 mm). Lightly water with a fine mist spray.

Cover with transparent polythene sheeting of 0.2 mm thickness mounted on the upper side of a square tight-fitting wooden frame of 5 cm thickness. Within 24 hours condensation of moisture will appear on the inside of the polythene cover. If moisture disappears at any time during the ensuing 7 days then give a light mist spray and close lid once more. Germination will usually begin by the 7th day and the cover can be removed. Keep sand moist. A seed is recorded as having germinated when it has reached a height of 1 cm complete with seed coat enclosing the cotyledons.

Record all germination from 7th to 28th day separately for each of the four boxes. Remove seedlings as soon as they are recorded plus any diseased ones which may not yet have reached 1 cm height as they may infect others. On the 28th day after sowing sieve the top layer of sand in each box through a No. 12 mesh screen (mesh size 12/64" or 4.76 mm) and record the number of full seeds which have not yet germinated.

The use of 4 replicates in a germination test enables the tester to measure the degree of variation in the sample. A simple method is to calculate the range of difference in germination % between the highest and the lowest of the sub-samples. Common causes of such situations are non-homogeneity of seed and, in nursery tests, fungal or insect damage in one or more of the replicates. Retesting is also necessary when an extremely high percentage of full, ungerminated seeds are left at the end of the test. In this case retesting should be done with a different pretreatment to try to overcome dormancy and improve total germination.

Purity and Germination Tests

Purity tests of most commercial seed lots of *Eucalyptus* are not made because it is difficult or impossible to separate the seed from the chaff of some species. Species in which seed and chaff are extremely similar in size, weight and colour include *E. cloeziana*, *E. regnans* and *E. delegatensis*. other small-seeded genera in which separation of pure seeds is difficult are *Alnus*, *Betula*, *Populus* and *Salix*. Even when separation is possible in these species, it is very time-consuming.

In the seed laboratory in Canberra it takes only 6–7 minutes to set up a test of eucalypt seed with 4 weighed replicates, but to carry out a purity separation and set up a 4 × 100 seed test takes from 20 to 50 minutes, depending on the degree of difficulty in separating out the chaff. The small size of seed also precludes a cutting test to determine the number of full but ungerminated seeds at the end of a test. For these reasons tests are best made on replicates by weight and results recorded as numbers of germinated seeds per unit weight of the impure mixture of seed and inert material. A squash test may also be used to give a rough estimate of viability.

Where seed is sown broadcast, the practising forester is primarily interested in the number of plants he can expect to get from a given weight of the seed lot he receives. Provided he is told that 1 kg of "impure" seed should produce 36,000 germinated seedlings, he is not worried whether this is caused by a combination of 90% purity × 80% pure seed germination or 80% purity × 90% pure seed germination. For local tests, therefore, if seed laboratory staff is insuffient, there is no objection to omitting the purity test even on species for which it would be practicable. In the case of direct sowing into individual containers, however, there are obvious advantages in having separate figures for purity and germination, since sowing is by numbers of (one to several) seeds per container rather than weight of seed per m^2 of bed or tray.

Viability Tests

Estimating the germination potential of a seed lot by actually germinating a sample of it is often the method most relevant to practical forestry. But the tests take several weeks to complete and for some species pretreatment may take some additional weeks or months. For this reason much research has been conducted to find other methods by which seed viability can be estimated accurately but much more rapidly than by germination testing. The following account of these methods closely follows that of Turnbull. The objects of quick viability tests are:

- to determine quickly the viability of seeds of species which normally germinate slowly or show dormancy under the normal germination methods;
- to determine the viability of samples which at the end of the germination test reveal a high percentage of fresh ungerminated or hard seeds.

Only two methods, the topographical tetrazolium test and the embryo excision test were previously accepted by the International Seed Testing Association as official methods for some species of seeds. ISTA has recently accepted the X-ray method as a valid alternative to the cutting test for the detection of empty and insect-damaged seeds. The following tests can be applied depending on the circumstances:

Cutting Test

The simplest viability testing method is direct eye inspection of seeds which have been cut open with a knife or scalpel. If the endosperm is of normal colour with a well developed embryo, the seed has a good chance of germinating. This test is not very reliable. Seeds with milky, unfirm, mouldy, decayed, shrivelled or rancid-smelling embryos and abortive seeds that have no embryo can be judged as non-viable without much difficulty. But it is not possible to distinguish moribund, recently dead or recently injured seeds which still appear the same as sound ones.

The cutting test, as already mentioned, is used at the end of a germination test to determine the apparent viability of ungerminated seeds; it is also a useful tool in estimating the size and maturity of the seed crop before collection and the efficiency of methods used in processing.

Tetrazolium Test

The tetrazolium method is only one of a number of biochemical tests which have been developed for seed testing. The various tests have been briefly reviewed by Moore. The tetrazolium test was introduced in 1942 by G. Lakon in Germany. In this method living cells are stained red by the reduction of a colourless tetrazolium salt to form a red formazan. The method emphasizes the need for a knowledge of the soundness of individual embryo parts for predicting the development of embryos into countable seedlings.

Normal practice is to soak the seeds in water for about 20 hours, then cut or puncture the seed coat to facilitate entry of the 1% aqueous solution of tetrazolium (TZ) and immerse the seeds in the dark for 48 hours. The process can be greatly speeded up by cutting through the seed at a distance of one third from the micropyle and placing it in a Vitascope vacuum machine for only half an hour. This method gives satisfactory results in Denmark but interpretation of results requires more experience in the operator than the method of seed immersion followed by excision of the stained embryo. The test is done on 4 replicates of 100 seeds each.

Justice states that, while the tetrazolium procedure is good in principle, its practical use in routine testing is limited by many problems, including: difficulty in staining of some seeds; necessity of cutting of dissecting seeds to permit observation of stained parts; poor agreement with results of germination tests in some cases, especially for seed of low germination capacity; lack of uniform interpretation of staining and difficulty in interpreting the significance of different degrees of staining; and an increase in man-hours required to test 400 seeds compared to regular germination tests.

Excised Embryo Test

By this method, the seeds are soaked for 1 - 4 days and the embryos are then excised from the seeds and placed on moist filter paper or blotter discs in petri dishes. The tests are placed in the light at a constant temperature of 20°C. The condition of the embryos is examined daily. Depending upon the species and lot differences, the tests can be terminated after only a few days, up to a maximum of 14 days, or as soon as distinct differentiation into viable and non-viable embryos can be made. The excised embryo test is similar to germination tests in that it measures the quality of the seed by their actual germination.

In addition it allows some measure of the embryo dormancy to be made, by counting those seeds which, although not growing normally, have grown slightly, remained firm and have kept their colour for the test period. The test is not valid for previously germinated seeds and must not be applied to samples which contain any dry germinated seeds. The success of the test requires considerable skill and experience in the operator and the ISTA rules restrict it to only a few species.

In a comprehensive study, Schubert compared the excised embryo method with the tetrazolium method for determining the viability of dormant tree seeds. He concluded that the tetrazolium method should receive preference over the embryo excision method but that improvements in the tetrazolium test should be made by providing for the use of bactericides and stronger reducing solutions to resolve doubts in weakly stained tissues.

Radiographic Methods

Radiography was first used to determine seed quality

over 70 years ago. The studies of Simak and Gustafsson highlighted the X-ray technique as a diagnostic method of tree seed analysis. The X-ray contrast method which uses various contrast or radiopaque agents was developed and applied successfully to species of *Pinus* and *Picea*. The X-ray method permits the detection of empty seeds, mechanical damage and abnormally developed internal seed structures, measurement of the thickness of the seedcoat and assessment of the seed viability when combined with a contrast agent.

The X-ray contrast method is based on the principle of semipermeability. When seeds are treated with a contrast agent, for example aqueous $BaCl_2$ or vaporous $CHC1_3$, their living tissues are able to prevent its entry due to their semi-permeability, but the dead tissues become impregnated. The impregnated tissues absorb X-radiation more intensively than the unimpregnated ones and thus appear lighter on the film than the unimpregnated ones.

The contrast permits living and dead tissue to be located in the seed and an estimation of its viability. There are now possibilities of using non-toxic water, instead of toxic $BaCl_2$ or $CHC1_3$, as a contrast agent for testing seed viability. The development of soft X-ray equipment has greatly simplified the operation. Complicated photographic equipment is not necessary and pictures can be made with polaroid film which provides clear and detailed radiographs within 30 seconds.

X-ray radiography has been successfully applied in determining the number of seeds in fruits of teak *(Tectona grandis)* and for studying their degrees of development. The technique has been tried on the fruits or seeds of sixty tropical forestry species and the results show that it can be reliably applied in processing such seeds.

A technique of stereoradiography as a supplement to the X-ray contrast method for use in seed quality testing has been developed by Kamra, Meyer and Wegelius. The chief advantage of stereoradiography is that it is possible for the observer to have a three-dimensional view of the object from a pair of radiographs. In this way, the exact topographical location of the contrast agent in the seed can be reliably determined. This increases the information which can be obtained from radiographs and adds to the analytical accuracy. The X-ray method is a useful one and is likely to play an increasing role in seed testing. Earlier models of X-ray machines were expensive, but recent models, particularly from Japan are much cheaper and now cost less than a cabinet germinator.

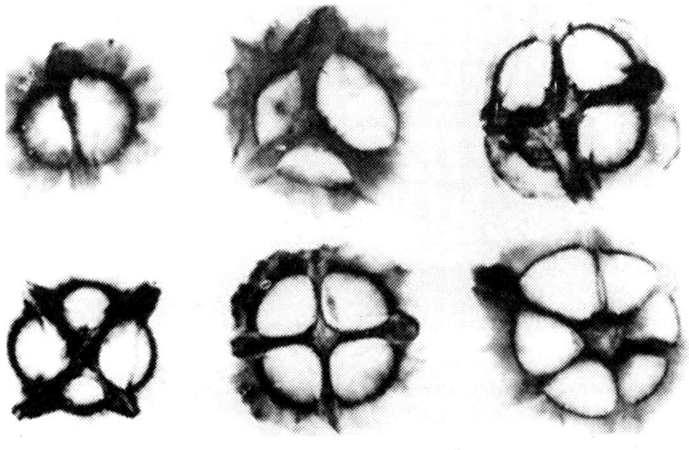

X-ray radiograph of teak fruits showing the variation in the number of locules (two to six).

Improvements in photographic films and paper have speeded up the process and simplified the interpretation, so that technicians can be trained easily to produce consistent results. ISTA has accepted the method as a valid alternative to the cutting test for the detection of empty and insectdamaged seeds. It also shows considerable promise for distinguishing between viable and non-viable seeds among "full seeds". For certain temperate conifers it has been possible to obtain good correlation between the Development Class (DC) of seeds, based on the development of both embryo and endosperm, and their germinability.

Hydrogen Peroxide

Hydrogen peroxide (H_2O_2) has a stimulating effect on seed germination and has been used in a rapid test for germination of several conifers in the western USA. Seeds are soaked overnight in 1% H_2O_2. The seedcoat is then cut open to expose the radicle tip and the seeds put back into 1% H_2O_2 in the dark at alternating temperatures (20° and 30°C). Counting and refreshment of H_2O_2 is done after 3 or 4 days and final assessment after 7 or 8 days.

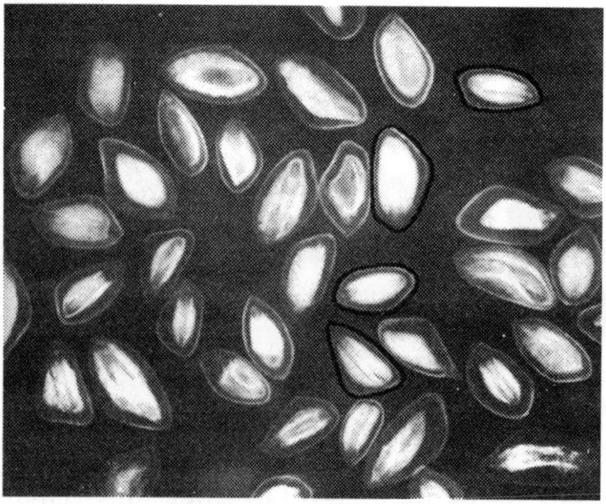

X-ray radiograph of Pinus caribaea seed.

Radicle growth of 5 mm or more is scored "evident", 0 – 5 mm "slight" and no growth means a nonviable or empty seed. The test is quicker but less reliable than a normal germination test (usually producing a more rapid and higher final germination), slower but simpler to perform than excised embryo and easier to interpret than TZ.

Moisture Content Testing

In order to control the operations of drying (or moistening) of seeds in preparation for storage and to check the stability of moisture content during storage, it is clearly essential to have reliable methods of measuring the amount of moisture in a given sample. Methods of determining moisture content of seeds have been classified by Justice into (a) basic methods in which the moisture is driven out of the seeds by heat and measured by the loss of weight of the original material, or the weight or volume of the condensed moisture, and (b) practical methods designed for rapid routine work and standardised against one or more of the basic methods.

Probably all the moisture cannot be driven out of seeds without driving out small amounts of other volatile constituents or causing chemical changes in the material which would result in weight changes. In applying any method, therefore, it is necessary to adhere closely to the prescribed procedure in order that the results of all tests made by that method will be comparable. Until recently ISTA prescribed three possible procedures:

1) Drying in an oven for 17 hours at 103°C
2) Drying in an oven for one to four hours at 130°C and
3) Toluene distillation.

Method (2) is applicable only to certain agricultural seeds and method (3), previously used for *Abies, Cedrus, Fagus,*

Picea, Pinus and *Tsuga,* has now been eliminated because it had ceased to be used in practice. This leaves only method (1) the "Low constant temperature oven method" as applicable to forest trees. The test should be made on two samples of about 5 g each, drawn from the working sample including impurities, not on pure seeds.

Large seeds should be ground, broken or cut into small fragments to facilitate drying and a good rule of thumb is that any seeds that average over 10 mm in diameter or length should be broken. The samples should be weighed and placed in metal containers, well-spaced to facilitate air circulation, within an oven which is maintained at a temperature of 103° ± 2°C for 17 ± 1 hours. At the end of that period the seed should be placed in a desiccator to cool for 30 – 45 minutes and then reweighed. The relative humidity in the laboratory where the final weighing is done should be less than 70%, to avoid rapid re-absorption of moisture.

The difference in MC of the two samples should not exceed a stated tolerance%. If it does, a further pair of samples should be tested; otherwise the mean of the two samples is the final result. The tolerance earlier prescribed by ISTA for all species was 0.2% but, as pointed out by Gordon and Bonner, a single tolerance figure is not applicable to all species.

The calculation of moisture content should be made on a wet weight or fresh weight basis i.e.

$$\text{Moisture content \%} = \frac{\text{Original weight} - \text{Oven weight}}{\text{original weight}} \times 100$$

Although wet weight basis is prescribed by ISTA and is becoming increasingly the standard form for expressing moisture content, it is not yet universal. To avoid any doubt, the method of calculating moisture content should be stated *explicitly* on any certificate or statement of results. As explained by Gordon and Rowe, provided that the initial fresh weight of a seed lot is measured and the initial moisture content (wet weight basis) calculated by oven-drying a sample, any new MC reached as a result of drying (or wetting) can be calculated directly from the new weight of the seed lot; there is no need for further oven-drying of samples at the new MC.

The desired weight of the seed lot to be achieved through drying (or wetting) can be calculated by multiplying its initial weight by the initial dry matter percentage and dividing by the desired dry matter percentage.e.g.

(1) If initial wet weight of a seed lot = 50 kg and the MC (wet weight basis), determined by oven-drying a sample, is 25%, the oven-dry weight = 75% of wet weight = 37.5 kg.

(2) If a period of drying reduces the wet weight to 46.5 kg, the new MC

$$\text{(wet weight basis)} = \frac{(46.5 - 37.5)}{46.5} = \frac{9}{46.5} = 19.4\%$$

(3) If it is desired to reduce the MC (wet weight basis) to 10%, then desired oven-dry weight will be 90% of new wet weight and the seed lot must be further dried until its wet weight

must be further dried until its wet =

$$\frac{50 \times 75}{90} = 41.67 \text{ kg.}$$

Electric moisture meters give rapid estimates of seed moisture but they are not considered accurate enough for official seed testing. Their rapid operation does make them very useful in certain situations, for example they should be accurate enough to check tree seed moisture as a guide to drying seeds for storage. Meter readings are converted to seed moisture content by means of charts supplied by the manufacturer or developed from calibration curves in the laboratory for the species in question.

Most meters will not measure moisture above 15 - 20% and require a minimum of 90 - 100 g of seeds for a test. A locally made, cheap and portable electric moisture meter has been used successfully in Thailand for several years to measure the MC of rice grains and could be used for tree seed of similar size. It measures electric capacitance and uses a 9-volt battery as the source of power.

Electric moisture meters are well suited to small seeds, but cannot be used for large seeds such as *Juglans* or *Quercus*; winged seeds such as *Fraxinus* are also difficult to measure. Large or winged seeds can be rapidly dried in a microwave oven. If the oven is preheated, drying can be completed in 5 minutes and weighing in 6 minutes immediately afterwards if in an electronic balance, or after 30 - 45 minutes' cooling in a desiccator if weighing is done in an ordinary balance. Results can be expected to be within 7% at a probability of 0.05 in the case of large seeds of high MC such as *Quercus*, and within 2% in *Fraxinus* and *Carya*, as compared with the more accurate results of the slower conventional methods.

A simple and cheap method of drying seeds quickly is to use an infra-red lamp. A weighed sample is subjected to heat from an infra-red lamp with such an intensity that it loses all its moisture, without being burnt, in about 20 minutes. When the loss of weight ceases, the new weight is measured and the percentage loss calculated.

2

Seed Processing

No matter how much trouble is taken, conditions for fruits and seeds are seldom ideal during transit from the forest to the seed processing depot. This is particularly the case if the journey is prolonged over several days. It is important therefore that fruits should be unloaded as soon as they arrive at the depot, inspected and placed in storage conditions which provide protection from rain, exclusion of rodents and birds and continuous free air circulation round the fruits.

Without such care fruits and seeds may suffer considerable deterioration between arrival at the depot and the extraction of seed. If moulds are allowed to develop on cones or fruits, they not only degrade the batch on which they occur but can form a source for infection of other batches by fungal spores. Because of the seasonal nature of most fruit and seed harvesting, large quantities of fruits are often collected within a short period. Seed extracting machinery has a limited capacity, so some of the fruits cannot be processed immediately on arrival at the depot.

Temporary storage is thus unavoidable. In some species it is also highly desirable, because it allows time for maturation and drying of seeds prior to further processing. Storage of cones or fruits for an extended period before extraction is necessary when an exceptionally heavy seed crop occurs, perhaps once in a decade or two, and very large quantities are collected in a single season. Research in British Columbia on 5 species of conifer showed that, provided good cone handling techniques were used, cones could be stored, either in covered outdoor cone sheds or in refrigerated coolers at 2°C, for six months from October to March.

In most cases germination was as good or better after six months than it was in the freshly collected seed. Controlled temperature storage offered no advantage over outdoor storage. In most cases, pre-cleaning of fruits is done soon after their arrival at the depot but, if the volume of fruits arriving at the same time is very high, pre-cleaning of part of the consignment may have to be deferred; in this case part of the storage period will precede the pre-cleaning operation and part will occur between pre-cleaning and extraction.

Fruits should be stored in a dry, cool, well-ventilated place to prevent moulding or heating. Depending on species, condition of fruits and processing techniques, sacks of fruits or cones may be emptied, loosely refilled and placed on storage racks; or the fruits may be spread in trays, on a storage floor or, under cover, on the ground. Floors of brick or wood are suitable, but fruits should not be placed directly on concrete floors because of moisture problems.

In Honduras slatted wooden storage bins supported at intervals on concrete blocks have proved very satisfactory for temporary storage of cones of *Pinus caribaea* and *P. oocarpa*. Gaps between the slats allow good ventilation and cones can be stored to a depth of 30–40 cm without the need to move them. In filling and stacking trays of cones, allowance must be made for a two- or three-fold expansion of cones as the scales open. Fruits may also be spread on tarpaulins.

Pre-Cleaning

Cones and fruits need to be cleaned of twigs, bark, foliage and other impurities before they go for extraction, cleaning, storage or sowing. In large extraction plants cleaning is accomplished with oscillating screens or vibrators. Precleaning by flotation is an alternative method. In small operations the major debris may be removed by hand. The impurities take up needless space. In addition leaf and twig fragments may carry fungal spores of e.g. needle cast disease of which the seeds themselves are free. These spores are a potential threat not so much to the seeds as to the newly germinated seedlings and to nursery stock and plantations in the vicinity of the nursery.

Impurities are more easily removed before than after extraction. Special measures are sometimes taken to remove sticky resin exudations from cones. Stein et al. reported that one company dried cones of *Pseudotsuga* sufficiently to harden the resin, then wetted and tumbled the closed cones to remove both dirt and resin. For some species, pre-cleaning, together with drying in some cases, is the only operation needed before storage or sowing. They are stored or sown as fruits. Pre-cleaning may include the removal of appendages to the fruit such as the involucre of *Quercus, Fagus* or *Tectona*. Winged fruits are often sown complete with wings e.g. *Ulmus, Fraxinus, Acer, Triplochiton, Pterocarpus,* several genera of dipterocarps.

Pre-Curing

Pre-curing is the deliberate storage and slow air drying of fruits and contained seeds in order to render them more suitable for subsequent operations of kiln drying, extraction and longterm seed storage. The processes assisted by precuring are seed maturation and fruit drying. Fruits do not all ripen at the same time, even for the same species and in the same forest. Thus, even when a collection is perfectly timed at the peak maturity of the crop, there will always be a proportion of sound seeds which are not fully mature. The minimum period for maturation in certain species is two weeks, but many demand more than 6 - 8 weeks.

Abies procera is one species which benefits from post-harvest storage of cones. It was found that during six weeks' storage, the seeds in the cones increased their dry weight by 10% as a result of accumulating organic materials from the cones; during this period carbohydrate and starch content in the seeds decreased but crude fat content steadily increased.

In a few species e.g. *Fraxinus excelsior, Magnolia* spp., *Gingko biloba,* the entire crop of seeds is dispersed with underdeveloped embryos which must complete their growth and development before the seeds can germinate. In most such cases of morphological dormancy, simple precuring is insufficient to induce germination, but warm, moist pretreatment is needed; this is often followed by a cold moist pretreatment to remove the physiological dormancy which also occurs in the same species.

A cool moist environment is best for artificial ripening and can be maintained by mixing the cones or fruits with peat moss or other inert moisture-retaining material in polyethylene bags. Satisfactory temperatures were 5°C for

Liquidambar and 17°C for *Pseudotsuga*. For *Pinus sylvestris* collection was done when specific gravity of the cones was 1.1 and the cones were stored for at least one month.

In New Zealand the correct sowing season for *Pinus radiata* is in October but ripe cones are not available for collection until November - December. The seeds must therefore be stored for nearly a year. Research has shown that immature green cones can be collected in June or July and artificially ripened by storing loosely in paper bags at a room temperature of 20°–24° C for 10 weeks. After kiln-drying and extraction, the germination of the seeds and the growth of the resulting seedlings were no different from those of seeds from mature cones collected in January.

Collection of immature cones can thus reduce the interval between pollination and sowing from three to two seasons, an important consideration when raising genetically improved stock from controlled pollinations. For most species of the humid tropics immature fruits can be encouraged to ripen by storage at room temperature in a sheltered, well-ventilated place. Temperatures below 20°C and above 35°C are likely to be detrimental.

Ventilation can be provided by keeping the fruits loosely packed in open bags or boxes so that normal respiration can take place. Rapid or excessive drying should be avoided. The aim should be to keep the fruits alive and healthy as long as possible so as to gain time for their seeds to mature. The fruits should be inspected daily and those that are ready should be removed for processing. Two categories of fruits require special attention, pulpy fruits and capsules. Pulpy fruits (drupes and berries) are ripe as soon as the pulp turns soft. After this point, the pulp begins to decay and ferment and this leads to seed deterioration.

Hence pulpy fruits that have become soft should have their seeds extracted as soon as possible. Capsules are ready when they split open of their own accord. Seeds forcibly extracted from unopened capsules are likely to be immature and non-viable. Precuring promotes a gradual decrease in moisture content of fruits (and seeds) which will shorten the period of kiln-drying necessary to make the fruits open. This saves time, power and money. It also prevents "case-hardening" of fruits, which can occur when fruits with a high moisture content are subjected to rapid drying and which makes subsequent extraction of the seeds very difficult.

Seed yield of *Pinus elliottii, P. taeda* and *P. palustris* was increased by 5 weeks of precuring before kiln drying. The effect was most marked in cones collected early in the season, as shown in the following table:

Number of seeds extracted per cone - Pinus elliottii

Date of collection	*Precured for 1 week*	*Precured for 5 weeks*
19 August	0	60
16 September	27	82

The germination of the extracted seed was also improved somewhat in *P. elliottii,* but not consistently in *P. taeda* or *P. palustris.* In Honduras the general rule for *P. caribaea* is to precure until all tissues have changed colour from green to brown. Promotion of free air circulation is of paramount importance, therefore fruits should be spread thinly (one or a few fruits deep) and should be regularly turned and stirred by raking.

Raised trays with fine wire-mesh bottoms are the ideal containers and, for best aeration and uniform curing and opening, a layer one cone deep is recommended. The fine wire-mesh will retain any seeds which may be released. In large processing depots it may be advantageous to raise the temperature gradually during the precuring operation. The final stage of precuring, before cones are passed into a kiln, can be carried out effectively by placing the cones near the top of the kiln where warm air from the exhaust passes through them.

Extraction Methods

The methods of extracting seeds from fruits are determined mainly by the characteristics of the fruits. Fleshy fruits are treated by a depulping process which usually involves a combination of soaking in water with pressure or gentle abrasion. Cones and other woody or leathery fruits are first dried until cone scales open or seeds become detached from the placenta of the fruit, and then treated manually or mechanically by tumbling or threshing in order to separate the dry seeds from the dry fruits.

Some indehiscent fruits, mainly nuts, achenes and winged samaras, do not require extraction but are stored or sown as fruits. Some species with seeds covered by a thin fleshy covering may be dried and sown with the dried skin intact. Drying under cover with frequent turning of the fruits is appropriate. Examples are *Vitex parviflora* in the Philippines, *Crataegus* in temperate regions and *Podocarpus* spp. and *Maesopsis eminii* in Africa. However, in some of these species germination may be improved by removal of the pulp.

According to their requirements for pre-storage handling and storage, the hardwood seeds can be classifieds in to three classes:

(1) seeds that must be dried for extraction and for storage;

(2) seeds that must be kept moist at all times, both during cleaning and during storage (i.e. recalcitrant species); and

(3) seeds that must be kept moist for extraction, then dried for storage.

Depulping

Depulping of fleshy fruits should be done soon after collection to avoid fermentation and heating. Small lots of seed are usually macerated by hand. After soaking, the flesh is hand squeezed or mashed by a wooden block, rolling pin or fruitpress. Alternatively flesh may be macerated by rubbing it against or through a screen. The pulp and skins can usually be separated from the seed by washing through appropriate sieves or by differential flotation in a deep bowl through which a slow stream of water is flowing. The seed sinks while the pulp rises to the surface.

In the Philippines the fleshy fruits of *Aleurites* spp., *Canarium ovatum, Syzygium cumini* and other species are placed in barrels or cans with water. After a day or two the pulp becomes soft. The fruits are then mashed carefully with a tamper without crushing the seeds. When plenty of water is added, the pulp will float while the seeds sink to the bottom. This method is also suitable for the fruits of *Gmelina arborea, Azadirachta indica, Ocotea usambarensis* and *Cinnamomum camphora* and for the syncarps or multiple fruits of *Chlorophora* and *Morus*. Because of the minute size of the seeds of *Anthocephalus chinensis* (2.6 million per kilo), a special technique is needed to extract them from the fleshy fruit.

Table 1. Some important hardwood genera classed according to the pre-storage handling and storage requirements of their seeds.

Dry for extraction and storage	*Always keep moist*	*moist for extraction and dry for storage*
1.	2.	3.
Acacia	Acer (some species)	Gmelina
Acer (some species)	Aesculus	Malus
Ailanthus	Castanea	Melia
Alnus	Corylus	Morus
Atriplex	Dipterocarpus	Nyssa
Betula	Hopea	Olea
Carpinus	Juglans	Prunus
CaryaQuercus	Rosa	
Casuarina		Sorbus
Cedrela		Ziziphus
Eucalyptus		
Fagus		
Fraxinus		
Gleditsia		
Liquidambar		
Liriodendron		
Nothofagus		
Platanus		
Populus		
Robinia		
Syringa		
Tectona		
Tilia		
Triplochiton		
Ulmus		

The outer portion of the fruit, which contains the seeds, is rubbed lightly against a ½ inch (12.5 mm) wire mesh. The mixture of pulp and seeds which passes through is placed on a screen box with a 1/16; inch (1.5 mm) mesh. The operators pour water over the mixture while rubbing it carefully by hand, so that the seeds together with some fine pulp pass the screen and drop into a water-filled container below. The seeds sink while the fine pulp remains afloat. If the floating pulp still contains seed, it is returned to the screen box and the operation is repeated.

In the Jari project in Brazil fruits of *Gmelina arborea* are depulped by macerating them against a wire mesh or mechanically by the use of a modified coffee depulper. Thorough cleaning of the stones is important. It has been found that thoroughly cleaned stones gave 10 % better germination than depulped but uncleaned stones. Fresh green fruits sown whole without depulping gave only 10 % germination while whole fruits sown after a period of drying failed to germinate. The stones may be cleaned in water or a combined cleaning and drying operation may be carried out in a rotating baffled steel cylinder which reduces the moisture content to 8 - 10 % after 20 hours at 45°C.

Removal of the pulp from drupes of *Vitex parviflora* in the Philippines has improved germination, both in full-size green fruits and in the riper purple fruits. The best improvement was from 26 % whole green fruits to 65 % depulped green fruits and the least from 38 % whole purple fruits to 52 % depulped purple fruits. Seeds may be washed free of flesh hydraulically. Fruit is placed in a mesh bag or wire basket and subjected to a stream of water from a high-pressure nozzle until all of the flesh and most of the skins are washed away.

After separation orthodox seeds should be carefully air dried under cover, with frequent turning. Thereafter they can go forward for shipment to the nurseries or for treatment to adjust moisture content to the correct value for the species, before storage. Where large quantities of fruits have to be depulped, various designs of machine are available. They include feed grinders, concrete mixers, hammer mills and macerators. Most machines only free seeds from the flesh; a part or all of the residue must be removed in later cleaning. The Dybvig separator, however, pulps the flesh and fully cleans the seeds in one operation. Small quantities of small-seeded fleshy fruits can be processed speedily by use of an electric mixer.

Fruits Drying

Drying, using either natural or artificial heat sources, must be used in the extraction of seeds of many important tree species and is almost always used for the cones of pines and other conifers and the capsules of eucalypts. Drying should imitate the natural drying process so that the fruits are subjected to progressive drying which causes a continuous release of moisture. Air coming in contact with the fruits must always be drier than the fruits and this can be obtained by continuous circulation of air.

Drying under Cover

This is the slowest and least severe method of drying fruits for seed extraction. The technique is the same as that described for precuring, but is employed as the sole method of drying rather than in combination with a later application of sun-heat or kiln-heat. Fruits must be in well ventilated rooms, spread thinly, stirred regularly if on a solid surface or, preferably, placed on trays with a wire mesh bottom to allow all-round air circulation.

Air-drying under cover is effective for cones of *Abies* and *Cedrus*, which readily disintegrate under this treatment and which can easily be damaged by heating in the sun or in kilns. It is also used for separating fruits of some hardwoods, such as *Quercus* and *Fagus*, from their involucral coverings. At the same time it is suitable for a modest degree of drying of these species and of others like them which need to be stored at a relatively high moisture content in order to retain viability. *Dipterocarpus, Hopea* and *Triplochiton* are tropical genera for which the method is suitable.

It can be used for drying the thin fleshy fruits covering of *Vitex, Maesopsis* and other species which are then stored or sown as dried fruits. The drying process is slow under this method and the period of time needed is dependent on the natural air humidity and temperature. But it is the safest method for any "delicate" species which will not stand heating or very rapid drying.

Sun Drying

This method is well suited to drying of cones and fruits of species which will withstand the rather high temperatures involved. It is commonly used in the dry season in tropical, sub-tropical or warm temperate climates where it can be 100 % effective in causing fruits to open and can make kilns unnecessary. In moist cool temperate climates it is much less reliable and may need to be supplemented, if not replaced, by kiln drying. Spreading the fruits in layers on screens, platforms, canvas or other sheeting in the sun is one of the simplest methods of air drying and requires little investment in equipment.

In the Mediterranean, pine species such as *P. pinea* and *P. halepensis* are handled by this method, as are *P. kesiya* and *P. merkusii* in Thailand and the Philippines. Fruits can be laid on wire screens with meshes of suitable size to let seeds drop through onto canvas or polyethylene sheets. The main requirements are:

1. Frequent stirring and turning to promote uniform drying and opening of the cones and release of the seed.

2. Arrangements for immediate covering of fruits in the event of rain, either by moving them indoors or by erecting a temporary shelter over them.
3. Care to avoid overheating of the fruits while they still have a high moisture content. This may involve preliminary precuring under cover or the avoidance, in the initial stages, of devices such as a base of corrugated iron sheeting or a covering of glass or polythene, which are designed to trap heat and raise temperature. The importance of this varies greatly according to the strength of sunlight locally and to the heat tolerance of individual species.
4. Frequent removal of any seeds which have separated from the fruits, prevent their being exposed too long to intense direct sunlight.
5. Protection against birds, rodents and insects which may pose a more serious threat in open air drying than in drying inside a building. Ants will carry off a large proportion of eucalypt seed unless they are rigorously excluded from the extraction area, while rodents and birds have a taste for the seed of pines.

Pines. Cones of Mediterranean pines exposed to sun drying take 3 to 10 days to open, depending on the drying conditions. In Thailand *Pinus kesiya* takes 5 - 7 days and *P. merkusii* 2 - 3 days. At Chiang Mai in Thailand, where the mean maximum temperature of the hottest month is 36.5°C, Bryndum found that putting the cones on a flat tray and then covering them with a clear polythene sheet, to produce a greenhouse effect with higher temperatures, gave nearly twice the seed yield in *P. kesiya* as the same treatment without the polythene sheeting; respective yields after 7 days were 15.6 g and

8.2 g of seed extracted per kg of cones. He also found that stirring the cones frequently (8 times a day) gave a significantly higher seed yield than a single stirring at the end of the drying period.

The trays used have 10 cm wooden sides with a bottom of 12.5 mm wire mesh. The seeds fall through the mesh into a sheet-metal funnel and from there they drop into a bag attached to the outlet. The trays rest on a wooden scaffold at a convenient height. A total tray area of 125 m2 is needed to extract 6 hectolitres of cones per day, and it was calculated that cost of materials and labour to construct it was one eighth of the cost of an imported electrical extractor of the same capacity. In addition to the tray method, locally made drums have also been used with success.

The material used is transparent roofing sheeting and there is an internal rotatable metal grid to hold the cones, which is rotated manually six times a day to ensure thorough mixing of the cones. Extraction period is similar to that in trays. Another method of multiplying sun heat which has been used effectively for *P. kesiya* in the Philippines is to place the cones on corrugated iron sheeting.

In Honduras *Pinus caribaea* cones are sun-dried either in trays or on tarpaulins. The trays are spread out individually in the sun but can be stacked and put under cover at night or in case of rain, or else the stacks may be covered with tarpaulins at those times. In the tarpaulin method heavyweight waterproof canvas is used, commonly of 5 × 7 or 5 × 10 m, but smaller sizes can be used if it is necessary to keep small seed lots separate (e.g. provenance collections). Cones are spread out one cone deep, leaving a perimeter strip of about 30 cm uncovered to avoid losses of cones by spilling off the tarpaulin.

During the day's exposure to sun the cones are moved every 2 - 3 hours by raking and lightly knocking the cones with the back of the rake. About 1 ½ hours before sunset or whenever it is about to rain, the two long sides of the tarpaulin are, in turn, pulled upwards and inwards abruptly until at least one third of the tarpaulin area on each side is free of cones. The cones finish up piled in a strip down the middle third of the tarpaulin. The pile is then covered by first lifting the short sides of the tarpaulin upwards and inwards by about 1 m, then the first long side (away from the prevailing wind), which is folded over the cones so that it just reaches the far side of the cone pile and finally the other long side which is pulled up and over the pile, on top of the first cover, ensuring that the edge reaches the ground.

It may be secured by weights or by tucking it under the pile. Extracted seed is best removed at the start of each day by uncovering the cones, tapping the cone pile with the back of a rake to dislodge the seeds, gently raking the cones towards the perimeter and then brushing the seeds together in the central strip and placing them in a suitable container. The tarpaulin should be laid on a well-drained site.

In Zimbabwe sun-drying of cones of *Pinus patula, P. elliottii* and *P. taeda* is done on a large scale in open-sided sheds covered with clear plastic roofs supported on wire netting. Each shed is 15.2 m long by 12.2 m wide and contains 8 wire netting troughs on a pole framework. Fresh cones are spread not more than two layers thick on hessian in the troughs. Those collected at the beginning of the season have a first spell of drying in the sheds followed by an intermediate period in half-filled sacks in racks in precuring sheds, followed by a final period of drying and cone opening in the troughs. The drier cones collected towards the end of the season need no precuring and open after a single short period in the drying troughs. The whole system of drying sheds has a capacity of 2000 bushels (720 hectolitres) of cones.

Eucalypts. Sun-drying is also effective for the capsules of *Eucalyptus* spp. wherever local climatic conditions are favourable. The following account is based on that of Turnbull. Small seed collections usually involve drying individual capsules or clippings of capsule-bearing branches. These can be spread out in a thin layer on canvas, calico or plastic sheeting, in a dry and well-ventilated position either in the sun or in shade. A small seed extractor can be made by placing the capsules on wire mesh a few centimetres from the bottom of a box and placing a sheet of clear plastic or glass over the top.

The capsules should be shaken daily and the seeds removed, so that they are not exposed for longer than necessary to high temperature. Bulk seed collections are often spread on tarpaulins on the ground, placed in special concrete enclosures, or raised up above the ground in frames on wires. Spreading the capsule bearing branches on the ground requires little equipment but is expensive because of the frequent turning required to permit the lower layers to dry out.

A more practical method of drying large quantities of capsules is to spread them over a frame covered with wire netting. The wire mesh permits a greater air circulation around the capsules than can be obtained if they are placed on the ground. A canvas sheet underneath the frame catches the seed as it is shed. Such structures may be large or small, permanent or temporary. Small temporary frames are suitable for use by the migratory seed collector and the large, permanent structures are placed in central seed extraction areas.

Another method sometimes used in drying is to hang capsule bearing branches over a single taut strand of wire. It has the advantages of providing very good air circulation between the leaves and the fruits and there is no risk of the material becoming compacted and heating up. The seed is collected on sheets placed under the branches. This method can be used for extraction in the forest, but can be equally well set up in a well-ventilated shed, the latter being used when there is a high risk of rain falling during the drying period.

A flexible system which is used in Brazil consists of placing the capsules in special trucks provided with shelves. Each truck has a capacity of about 50 kg of capsules. During the day the trucks are left in the sun and at night or during rain they can be wheeled under cover. The fruits usually open in about three days. The rate of seed release during natural drying varies with the capsule characteristics of each species, capsule maturity, and most importantly, on the drying conditions.

Very mature capsules of some species can release their seed within a few hours under optimum drying conditions but under average conditions most species will dry sufficiently in 3 - 4 days. Some eucalypts characteristically retain unopened capsules on the tree for several years, these capsules may become very woody and are often difficult to open. Although drying in full sun leads to rapid opening, there is an inherent danger that, if the temperature becomes too high, any primary dormancy in the seed may be strengthened. The capsules should not be placed directly on metal which is exposed to the sun, since this may result in high temperatures which damage the seed.

During natural drying ants and birds can cause considerable seed losses. Ants will carry away viable seeds and leave the chaff behind. Spraying an aerosol insect repellant or spreading an insecticide dust around the seed bearing material will usually discourage pilfering. Seed-eating birds such as the domestic sparrow can also be responsible for seed losses during extraction. *Other species* which may be sun-dried include dry zone legumes such as *Acacia* and *Prosopis* spp. and various species of *Toona, Lagerstroemia, Leucaena, Casuarina, Albizzia falcataria* and *Pithecellobium dulce.*

In India the "cones" of *Casuarina equisetifolia* are placed in trays in the sun and a thin cloth is secured to the top of the trays to prevent the "seeds" from being blown away by wind. The "cones" are treated with BHC 10 % powder or other insect repellant to prevent ants from removing the "seeds". In three days the "seeds" separate from the "cones", if left longer other "cone" fractions get mixed with the "seeds". In general, hardwood fruits and seeds are more easily damaged by overheating than those of conifers, so care must be taken to avoid prolonged exposure of the seeds to intense direct sunlight. Intermittent shade may be necessary.

Artifical Drying

Drying of fruits in heated kilns may be necessary for a variety of species where the climate is not suited to air drying i.e. in cool moist climates. It may also be necessary for a few refractory species which will not respond to sun-drying even in a dry climate. It is used most often for the cones of coniferous species, but also for some eucalypts in cool moist areas. Although seed of most species of *Casuarina* may be readily extracted by sun-drying, some species have serotinous cones which require the higher temperatures of a kiln (or, in nature, bush fires) before they open.

Kiln drying is usually preceded by a period of air-drying or precuring. The main difficulty in drying cones by natural methods is the impossibility of controlling air humidity and temperature. An increase in the humidity of the air may cause a reclosing of the cones Artificial heating, in contrast, permits control of air moisture and temperature, a much shorter period of treatment and, with a continuous process, more effective work organisation. Artificial heating requires expensive equipment and installations, which are not used throughout the year. This makes their unit cost exceptionally high.

A very careful appraisal must therefore be made of the capital cost, in relation to the amounts of seed to be processed annually, before establishing large, permanent kilns. As much use as possible should be made of air drying. Sometimes a combination of the two methods is appropriate; a small kiln can be installed in order to supplement air drying of some species by a final short period of kiln drying, or to deal with species or batches of cones which prove resistant to air drying alone.

Cone drying by artificial heat should be carried out in such a way as to obtain drying in the shortest possible time without damaging the viability of the seed. For this purpose the following recommendations are applicable:

1. Cones should be properly precured before they are put in the kiln.
2. Air temperature should be controlled and kept to the minimum level sufficient to dry cones.
3. Cones should not be heated nor seeds kept in the kiln longer than necessary.
4. Air in the kiln should be kept as dry as possible.

The principle of kiln drying is the provision of a regulated flow of warm dry air, so that all the cones dry uniformly and as quickly as possible without risk of overheating or "case-hardening". In most modern kilns the temperature of the air-stream is raised progressively as drying proceeds; there is forced-draught air circulation and seeds are speedily removed from the source of heat as soon as they are freed from the cones.

Cone kilns vary in size from small cabinets to large buildings, but they have in common a source of heat, a means of controlling movement of heated air by forced-draught or convection and some tray, shelf or other system of exposing cones to the moving air. Kiln controls may be simple and hand-operated or complex and automated. Some kilns also have sensitive humidity controls.

In many kiln schedules the first two or three hours of drying is at a considerably lower temperature than the effective drying temperature. This prevents the combination of high temperature and high moisture which is the most common cause of reduced seed viability during drying. The safe temperature of all cones is about 30°C, rising to 60°C when the moisture content of cones is below 10 %. Humidity control is available in some modern kilns but optimum schedules have yet to be determined for many species.

Types of Kiln

Turnbull classified kilns under the following types (a) Stationary tray kilns (b) Vertical progressive kilns (c) Horizontal progressive kilns (d) Rotating drum kilns (e) Portable kilns.

Stationary tray kilns

The simplest kilns are convection kilns consisting of

anything from a stove-heated room to a more complex structure. Basically they consist of a heating unit above which is an extraction chamber. Heat enters the chamber at the base and passes up through tiers of trays containing the fruits or cones. Since the air becomes cooler and moister as it rises through the cones, the efficiency of extraction is not uniform in all parts of the kiln. Some trays must often be moved to different levels and re-treated. Such kilns are relatively cheap and they require no special technical skills to operate. They have the disadvantages that they are relatively inefficient, fire risk is often high, and temperature and humidity control are rarely adequate. They can be improved to some extent by the provision of forced ventilation.

In Zimbabwe most cones are air-dried, but this is supplemented by a simple kiln using tobacco-curing flues. Because there is no mains electricity in the area air circulation is by convection and the kiln is wood-fueled. Safe maximum temperatures are 48° C for *Pinus elliottii, P. taeda* and *P. kesiya* and 60° C for *P. patula*. The kiln holds 36 trays, equivalent to 9 bags (about 22 bushels or 8 hectolitres) of cones, which will open in one shift of 8 hours.

In Honduras cones of *Pinus oocarpa* and *P. caribaea*, after a period of precuring, are successfully dried in a forced draught hot air kiln modified from a solar kiln for drying timber. Ventilation is provided by two large diameter fans powered by a one horse-power electric motor, while heating is provided by flue radiator and a furnace using waste cones and firewood. Air can be recirculated if required by opening a valve which connects the air entry and exit passages.

The kiln holds 32 hectolitres of closed cones, arranged in 8 stacks of 8 trays each, with 50 litres of cones per tray.

The temperature regime is 4 hours at 40°C followed by 10–14 hours at 45°C for *P. caribaea*, while the thicker cones of *P. oocarpa* receive a constant 50°C. Cone opening is usually complete in 12–18 hours. Seed yield is 160–250 g per 100 litres of closed cones for *P. oocarpa*, 125–625 g per 100 litres for *P. caribaea*.

Electric ovens, preferably of forced draught type, are useful for drying small lots of eucalypt capsules. In Brazil electric ovens, consisting of a large chamber with a series of sliding shelves and a capacity of 80 kg of material, are used for drying eucalypt capsules. Drying time is 24 - 36 hours at 45° C. A forced draught oven has been used sucessfully in the USA to dry pods of *Prosopis*. The pods are placed in the oven at 32° C for 18 hours. The dried pods are then placed in an electric scarifier for 10 - 15 seconds, after which the seeds are separated by removing light debris in an air-column blower and sieving out the pod fragments with a size 11 or 12 screen (mesh size 1.85 or 1.70 mm).

In Sabah pods of *Acacia mangium* are dried in a simple drying chamber which incorporates an electric heater and a domestic fan. Pods are held in stacks of trays 0.7 × 0.7 m square, each with a wire mesh base. Air entry temperature is 45°C. Pods are separated by colour and treated as follows:

(1) Black pods - no precuring, dried for 24 hours in drying chamber

(2) Brown pods - precure in shade for 48–72 hours, then dry in chamber for 24–48 hours

(3) Green but full-sized pods - precure in shade for 72–120 hours, dry in chamber for 48 hours. The same method is used for *Albizzia falcataria*.

In place of trays, steel-mesh cylindrical drums are sometimes used to hold the seed-bearing branch clippings of eucalypts. Kilns of this type are in use at more than one site in Tasmania. One kiln is approximately 9 m × 4 m and has an extraction capacity of about 4500 kg of seed per annum. It has insulated walls with a transparent roof, formed of two layers of acrylic sheeting with an air space between, which permits solar heating in summer.

In winter a false reflective ceiling is put in to reduce loss of radiant heat. The kiln is maintained at a temperature of 40° C. After 36 hours the capsules have opened and the cylinders are rotated and shaken to dislodge the seeds. If no special equipment is available and only a few fruits or cones are to be dried, they may be placed on a warm radiator. They should never be left to open where it is too hot to rest the hand comfortably.

Vertical progressive kilns

The cones are placed in the kiln on a set of trays, one above the other, in a tower-like structure; trays descend during cone treatment. From the bottom of the kiln the hot dry air hits the lowest tray, containing fairly dry cones; passing through the cones, the air loses some heat and takes some moisture from the cones. In the second tray from the bottom, cones have somewhat higher moisture content and the air is less hot and less dry. In the topmost tray, the cones, still having the original moisture content, are engulfed by warm, humid air.

At the right moment, a tray is taken away at the bottom, and the whole tray set descends one step, while a new tray with fresh cones is put on the top of the pile. Hot air can circulate simply by convection, but forced ventilation makes the operation quicker and more regular. The distance between two successive trays should be sufficient to contain open cones which have a volume 2 to 3 times that of closed cones.

The time required to complete drying varies according to the initial moisture content of the cones, the volume of circulating air, and the type of air circulation (convection or forced ventilation). In some kilns, cones are forced to fall down from one tray to the next below by opening the tray bottom, or by inclining the tray. In other types, trays are replaced by slow moving bands; cones move on a band till they fall to the one below. In the last two types, cones are shaken as they fall from tray to tray, so that most of the seeds come out: special devices under the bottom tray collect the seed, while cones pass to further handling.

Horizontal progressive kilns

The operation of this type of kiln can be illustrated by the Italian designed horizontal kiln described by Gradi. A continuous conveyor belt of perforated steel plate on which the cones are spread passes through an insulated galvanised tunnel. By means of radiators and ventilators installed in chambers on one side of the horizontal belt, an air movement in the form of a spiral is obtained from the entrance to the exit of the drying tunnel. In the first few metres of travel the cones undergo a pre-drying stage at a temperature about 10°C lower than the final temperature.

In successive sectors of the drying tunnel higher temperatures are used, usually in the range of 45°–50°C. The spiral movement of the air creates a turbulence which greatly accelerates the drying process. The kiln described by Gradi is designed to process large quantities of cones. More versatile equipment has been produced to process and maintain the identity of small seed lots as well as large ones. This comprises a long horizontal kiln having small bins with removable trays rather than a conveyor belt.

The kiln is provided with portable heaters which are rotated each day when unloading the kiln to ensure that the hottest air is on the driest cones. The static pressure and air flow is controlled through each bin. This equipment has a particular advantage in being very easy to clean. An alternative simple method of preserving the identity of small seed lots is to use small nylon bags. They allow free air circulation and the same bag can be used for the cones from the time of their insertion in the forest throughout transport and extraction. The advantages of the horizontal kiln over the vertical kiln are

(1) the expensive process of elevating cones to the entrance of the conventional vertical drier is avoided;

(2) any possibility of damage to the cones in falling from one level to the next is avoided, because they remain stationary on the conveyor during all phases of the drying process;

(3) the high productive capacity;

(4) the ease of installation, inspection and maintenance.

The relatively high cost of the equipment requires a high through-put of cones for optimum economy.

Rotating drum kilns

This type of kiln can be found in many modern seed extraction plants. The basic operation is as follows:

Cones are put in a cylinder made of perforated steel plate, rotating on a central axis. The cylinder is contained in a box, where forced air circulation is provided by an electric fan. During the extraction the cylinder rotates continuously, shaking the cones. The air temperature increases gradually, from room temperature to the maximum fixed level. The strong ventilation, the progressive heating and the continuous shaking make the cones dry and open in a short time. Seeds released from cones pass through the holes of the plate, and are blown at once out of the kiln. Temperature and ventilation control is generally fully automatic.

Seed is taken out of the hot air as soon as it comes out of the cone, in order to avoid seed damage. The capacity of the cylinder is reduced so that extraction can be done separately for cones from different provenances, even if the quantity is small. Kilns of this type may have electrical heating or a separate heating unit which can use different fuels. The cylinder, the electric motor, the ventilator and the equipment for air and temperature control are contained in a steel box, above which is a pre-drying box with two floors. From this box, cones fall into the cylinder, where they are exposed to gradually increasing temperatures (40–45–50–60°C) for three to four hours.

At the end of the extraction, reversing makes the cylinder open, the open cones fall down and other cones come into the kiln. Modern rotating drum kilns are generally of compact metal construction and of moderate size so that they can be placed in small, inexpensive buildings. Their output is relatively low and they are generally planned to work in a set of two or three depending on the requirements of the extractory. Their advantages include the relatively short drying time and the shaking of cones during drying which extracts the seed and makes a separate tumbling operation unnecessary. This type of kiln is widely used for modern seed extraction of conifers and is being used for the extraction of eucalypt seed in Australia.

Portable kilns

For small-scale extraction or to provide seeds for modest

plantation programmes or for research, large commercial cone drying kilns are neither satisfactory nor economical. In research work it is often vital that seed be extracted from small lots of cones without the possibility of the seeds being mixed with other lots. A kiln designed to dry small lots of pine cones is described by McConnell. It features economy, safety, portability, versatility and a small load size.

The kiln consists of a heating mechanism based on bottled gas, a fan, a control panel, and both a drawer type chamber and a cabinet chamber to hold the cones. The capacity is 50 bushels (18 hectolitres or 1.8 m^3). The control panel is designed to sound an alarm or provide a visual warning of any malfunction in the equipment, and thermostats ensure that the temperature cannot, under any circumstances, exceed 74°C. The portable kiln described above could find wide application for small operations through its economical operation and advanced safety features.

Safety Precautions

Any artificial heating involves a fire hazard and this is particularly true in handling cones, since dust, resin and dry cone scales are all highly inflammable. Stringent fire precautions including a ban on smoking should be enforced, fireproof, non-wooden construction materials should be used and arrangements made for frequent removal of inflammable dust and debris by vacuum equipment or other means. Other precautions are necessary when drying certain dry fruits and seeds of certain species. Dust masks need to be worn when handling species such as *Platanus* spp. which release fine hairs which might be breathed into the lungs.

Separation

When fruits and cones open after drying, some seeds fall out easily as a result of manual stirring, rotation in rotating drum kilns or, in certain vertical progressive kilns, from the shaking of the cones as they fall from one tray to another. But many seeds are left inside, especially in those drying techniques where the cones remain static. They must be removed as soon as possible after drying is complete.

In some species a thorough manual shaking is sufficient to extract the remaining seed. The capsules of eucalypts need to be vigorously shaken, particularly if not fully mature, because abscission of the seeds from the placenta may be only partially complete. Failure to shake slightly immature capsules properly can result in only the chaff being released. The fertile seeds are usually attached to the placenta near the bottom of the loculus so that, after dispersal of the chaff, immature capsules may, on superficial examination, appear to be empty. Cones may be shaken in coarse sieves to release the seeds, but more vigorous treatments are needed for some species. Those most widely used are tumbling for conifers and threshing for hardwoods.

Tumbling

A tumbler is a rectangular or round container or drum mounted horizontally on its long axis. As it turns, cones tumble about; interior baffles often accentuate the jarring and tumbling action. Seeds fall from open cones through the high-strength wire mesh which forms the sides of the tumbler and into a hopper or trays or onto a moving belt. The tumbler can be operated by hand or mechanically driven, depending on the scale of the operation. Some drums may be closed at both ends and emptied and refilled at the end of each operation cycle.

In more modern designs a continuous operation can be achieved with an inclined cylinder open at both ends. The cones are fed in at one end and during rotation roll slowly to the other end where they are discharged. Variable speeds and tilt are set for each species. The speed determines the rolling and pitching effect on the cones, while the tilt determines the length of time the cones remain in the tumbler. Small types of tumbler are easily transportable. Fisher and Widmoyer describe a small tumbler of ½ bushel (18 litre) cone capacity made from a modified domestic washing machine.

In Zimbabwe a hand-operated drum tumbler 2.43 m long is used, fed from a chute from the floor above. By building the tumbling room on a slope, the cones can be easily transported into the upper floor. The drum holds one bag of cones and it takes one minute to tumble one charge and to recharge the drum. The seeds drop through the 18 mm mesh of the drum into a collecting tray below and the empty cones are discharged into a trolley. It is important that tumbling should be carried out as soon as possible after drying, because open cones exposed to cold wet air can reclose in a short time.

If tumbling cannot follow immediately after drying, the opened cones should be stored in warm, dry conditions in the interval between these operations. The time required for tumbling depends on the species and the condition of the particular lot of cones being handled. Seed in some species such as *Larix decidua* and *Picea abies* may be held tightly in the cone and long periods of tumbling are sometimes required to extract them. Special machines, such as a large potato-peeling machine or a tumbler incorporating saw blades are effective on difficult species like these.

Alternatively, the cones may be rewetted and then redried to promote fuller opening of the cone scales.

Haverbeke found that, after the first tumbling in *Pinus sylvestris*, soaking the cones in trays of water at 30° C for about half an hour until the cones softened and started to close, followed by thorough air-drying until the cone scales opened again, gave good results. The yield of seeds from the second tumbling averaged 36 % of that from the first. There was a marked difference between provenances, from 18 % for a Scottish to 84 % for a Spanish origin. The additional yield from the second tumbling fully justifies the extra cost for rare and valuable seeds such as those obtained from controlled pollination.

Mechanical damage can easily be inflicted on seeds if excessive tumbling speeds are used or if the tumbler is filled with too many cones. Speed of rotation and time of treatment should be adapted to the cone and seed characteristics of the species being handled. It is better to leave some seed in the cones than to spend money on extracting seeds most of which will be severely damaged in the process.

Threshing

Extraction of seed from dry fruits of many hardwood species is accomplished by threshing. Seed extraction in species such as *Cercis, Catalpa, Robinia* and *Liriodendron* is easily accomplished by spreading the fruits on a platform, sometimes on a straw mat or other suitable material, and beating them with a flail or slender pole. For large quantities mechanical threshers used in agriculture can be adjusted for tree fruits by altering the distance between the crushers. In chile pods of *Prosopis tamarugo* are ground in a stone mill set at 4 mm and the seeds are then separated by sieving and floating the milled product.

A modified cereal huller has been used as a thresher for *Prosopis* pods in the USA. The machine, described in

Ffolliot and Thames, will thresh 1 bushel (36 litres) of pods in 1 ½ hours; approximately 160 hours would be required to thresh the same quantity by hand. In the Philippines fruits which do not readily release their seeds are put in a sack and beaten. Separation is done by use of screens. For every seed size one screen is used with a mesh larger than the seeds, to separate them from fruit fragments and other large impurities, and another with a smaller mesh which retains the seeds and allows the fine impurities to pass through.

In Sabah seeds of *Acacia mangium* are separated from the pods, after drying, by rotating them for 10–15 minutes in a cement mixer together with blocks of hard timber 10 × 10 × 15 cm. A similar use is made of a cement mixer for pods of *Albizzia falcataria* but, because separation of the seeds is easier in this species, it is not necessary to include the timber blocks. Several types of mechanical thresher suitable for *Acacia* pods are described by Doran et al.; they include a hand type model of resilient tapered thresher, a rotating drum, a flailing thresher and a peg drum thresher.

Many acacias give off a very irritating dust during threshing and protective equipment should be worn by operators. More robust methods, such as pounding the fruits with a wooden pestle or putting them through hammermills, must sometimes be applied. For some species special equipment has been developed such as the dehuller of *Juglans*. Hammermills consist of a hooded inlet or hopper, a central chamber containing a series of hammers which rotate about a central shaft, and removable outlet screens of different mesh.

The outlet screen must have holes large enough to let seeds pass through without damage. Fruits are fed into the mill continuously during separation. Care must be taken to operate hammermills at comparatively low speeds, 250 - 800 revolutions per minute, to avoid injury to the seeds. The Dybvig separator works well on dry as well as on fleshy fruits.

Other Extraction Methods

Seed extraction of some species is difficult, even after standard treatments of drying and tumbling or threshing. In the Philippines indehiscent hard pods of leguminous species such as *Delonix regia, Pithecellobium saman, Cassia fistula, C. javanica* and *Parkia javanica* have to be opened with a machete or knife and the seeds picked out individually. The pods of *P. saman* are sweet and relished by termites; if they are piled in a dark place, after some time only the clean seeds are left behind.

Goor and Barney recommend that *Cedrus* seed be stored in the cones, since extracted seeds quickly lose their viability. When removed from storage, the cones must be soaked in water to facilitate extraction. After treatment the cones can be easily broken apart by hand and the seeds extracted for immediate sowing. Serotinous cones of species such as *Pinus brutia, P. halepensis, P. contorta* and *P. radiata* may need special treatment to induce opening. Dipping them in boiling water for 10 to 120 seconds (up to 10 minutes for some especially refractory seed lots), followed by very high temperatures in the drying kiln (75° - 80° C), has been effective in some cases.

High temperature is needed to melt the resin which forms a strong bonding adhesive between overlapping cone scales. Immature green cones may also need special treatment. It was found that green cones of *Pinus merkusii* from Zambales (Philippines), after soaking for 48 hours, followed by 80 hours' drying at a starting temperature of

30° C and a final temperature of 50° C, released only 7 % of the contained seeds; if the cycle of alternate soaking and drying was repeated 5 or 6 times, 79 % of seeds were released.

The total period of 4 - 5 weeks would be uneconomic in operational forestry. The recommended alternative is to collect only ripe brown cones which released 91 % of their seeds in a single cycle. When seeds have been extracted from their fruits, several operations are needed before they are fit to go into storage. Sound seeds must be separated from empty and non-viable seeds and from inert fragments of fruits; winged seeds of some but not all species need to be dewinged; if seeds are to be stored, their moisture content must be tested and, if necessary, raised or lowered to the percentage most suitable for storage.

If uniformity in growth of nursery stock is considered desirable, seeds may also be graded by size. Inert material takes up space both in storage and transport and may cause uneven stocking in the nursery seed beds. It also carries a greater risk of introducing pests or diseases than do the seeds themselves; for example spores of needle cast are carried on needle fragments rather than seeds. Cleaning to a high standard of purity is easy in some species, but more difficult in others.

Cleaning of seeds to a purity higher than a given percentage is undesirable in some species; beyond that, an increasing amount of good seed is separated out with the impurities. Besides, the added effort of extra cleaning is time-consuming and expensive. Morandini recommended that seeds of *Larix* should not be cleaned to a purity higher than 65 %, since it has been shown that further cleaning causes a drastic loss of good seed. This is because the great thickness of the seedcoat in comparison with total seed size makes empty seeds nearly as heavy as full seeds.

In many species of eucalypt, especially in the subgenera *Monocalyptus* and *Idiogenes*, it is difficult to separate the fertile seeds from the chaff which is released from the capsules at the same time. Viable seeds are frequently very similar in size, shape and colour to the chaff particles and the proportion by weight of chaff to viable seeds is usually in the range of from 5 : 1 to 30 : 1. Commercial seed lots of eucalypts are therefore cleaned of leaves, twigs and other large fragments but the remaining "seed" consists of a mixture of seed and chaff; this practice is accepted by both buyer and seller.

Provided that seed lots are accompanied by test results giving the number of viable seeds per unit weight of seed plus chaff, the seed user will not be too concerned that they contain a certain amount of impurities. Cleaning seeds for operational use should therefore be applied with discretion. It may be necessary when special techniques are used e.g. if the seeds are to be pelletted or if precision sowing is to be employed. For research purposes also, when basic knowledge on germination or other seed characteristics is sought, repeated cleaning to a very high standard is necessary.

Dewinging

Winged seeds or winged fruits are a feature of many forest trees and almost all conifer seeds have a wing which may vary from long and hard to very short and soft. In order to make seed processing and nursery sowing easier, the wing is usually removed whenever it is larger than the seed (or fruit). Wings are small or impractical to remove from seeds in several coniferous genera e.g. *Thuja, Chamaecyparis, Cupressus;* in a few genera they cannot be removed without impairing seed viability e.g. *Libocedrus*.

Many winged hardwood fruits e.g. *Casuarina, Betula, Ulmus* are stored and sown intact, but the larger wings of e.g. *Swietenia* fruits may be broken off. For small quantities of seed dewinging may be done by hand, either by rubbing seeds between the hands or against a screen or roughened surface, or by handrubbing in a cloth bag or by rolling them between two cloth sheets or in a cloth bag between a rubber surface below and a roller above. For large quantities mechanical dewinging is in common use.

Dewinging machines range from those which are hand operated to large semiautomatic equipment which gives a continuous output. Corn mixers and cement mixers are frequently used. Mechanical dewinging, if carelessly done, may cause damage to seeds by crushing, cracking or abrasion. Most mechanical dewingers are rotating devices in which the seed is pressed by brushes or pads against the walls of a cylinder, or rotating knobs or pads force the seeds to pass through narrow apertures while retaining the wings.

If the clearance between the knobs or brushes is too small the seed may be damaged. A dewinger specially designed for small seed lots of 5 kg or less and for easy cleaning between seed lots has been described by Lowman and Casavan. It consists essentially of a rubber-lined cylinder with a rotating central shaft to which are attached pure gum rubber flaps. There is a variable inclination setting so that seeds and wings will pass out of the dewinger by gravity.

Nartov et al. describe a combined dewinging and seed cleaning machine in use in the USSR. It is transportable and weighs 50 - 70 kg. A seed hopper and feed auger pass seeds into the dewinging unit which has brush-type beater vanes. A fan then removes wings and lighter material, while the heavier material falls onto a series of inclined screens with various mesh sizes. The clean seeds fall into different containers, graded according to seed size.

Mechanical injury can be avoided in some cases by moist dewinging. Wang describes a safe method of dewinging coniferous seeds in Canada. The seeds are moistened with water and left to soak for 20–30 minutes before they are stirred with a soft brush or sponge in a rotating cement mixer to remove the wings. A similar principle is used by Isaacs in dewinging pine seeds. A large tank with slow-moving pipes gently agitates the seeds, which have been moistened at a rate of about 2 litres of water to 45 kg of seed. The wing absorbs moisture and is shed from the seed.

In Honduras moist dewinging of *Pinus caribaea* and *P. oocarpa* is done through the use of a small cement mixer or rotating drum. The capacity of the drum should be at least double the amount of seed to be dewinged and the speed of rotation about 1 revolution every 2–3 seconds. The winged seeds are tumbled dry in the drum for 15 minutes; water is then sprinkled slowly over the seeds as evenly as possible while the drum continues to rotate, at a rate of about 1 litre to every 50 litres of seed.

Tumbling is continued for 45 minutes after adding the water, and then the mixture is emptied out on to a gauze-bottomed tray, after which the seeds are separated from the wings. Moist dewinging is also in common use in Sweden. In moist dewinging the seeds absorb water and have subsequently to be dried to an acceptable moisture content.

Seed Cleaning Methods

The main characteristics by which sound seeds may be distinguished from inert matter including sterile and empty seeds are size and shape, specific gravity, colour and surface texture. The ease with which sound seeds can be differentiated depends on (1) the degree of difference which exists between the seeds and the matter to be separated from them and (2) the degree of uniformity among the seeds themselves. Colour, size and shape are useful criteria for visual separation, while most seed cleaning machines make use of seed size and specific gravity.

Screening and sieving methods separate by seed or particle thickness or diameter; the indented centrifugal cylinder by particle length; liquid flotation and blowing, fanning and winnowing methods by specific gravity; while frictional cleaning methods rely on differences in surface texture. Modern cleaning machines often combine more than one method, so that the cleaning process is both effective and quick. However, the species and the amount of seed to be handled will determine whether cleaning is best carried out by hand, by improvised equipment or by specialist machinery. Some cleaning and grading methods are described below.

Screening

In most cases a number of sieves with different sized perforations are used and the cleaning is a process of gradually sifting out smaller and smaller particles. It is not only the size of the perforations which determines the quality and quantity of the seed cleaned; other important factors include the precision of the perforations, the angle at which the sieves operate, the amplitude and speed of movement of the sieves, and the correct cleaning and maintenance of the equipment.

Sieves or screens may be made of flat perforated plate or wire mesh, and sometimes they may be three dimensional such as funnelshaped sieves. For small samples hand held sieves are adequate, but in larger scale cleaning a series of shaking screens is commonly used. In Brazil sieving with a mesh size of 32 per inch (approx. 12.5 per cm) was effective in separating seeds from chaff in *Eucalyptus grandis*. 84% of good seeds were retained and 89% of chaff was eliminated.

Sorting According to Length

Cleaning with sieves relies on the separation of seeds, with diameter the critical factor. Fractionating according to length cannot be done with sieves, but this is possible with an indented cylinder. In addition to its use for separating good seed from the impurities, the equipment is used in agriculture for separating seed mixtures and can also be used for grading seeds.

The equipment consists of a slightly inclined horizontal rotating cylinder and a movable separating trough. The inside surface has small closely spaced hemispherical indentations. Small material is pressed into the indents by centrifugal force and can be removed. The larger material flows in the centre of the cylinder and is discharged by gravity. Depending on the type of impurities, the seed may be separated via the indentations or by passing down the cylinder.

Blowing

Cleaning by blowing is a very important and widely used method. It is based on the principle that any object can float in an airstream of sufficient velocity. For separation in an airstream there are three possibilities: falling, floating or rising. The behaviour of the seed and other matter will depend on their weight, their resistance to the flow of air (volume and shape), and the velocity with which the air moves.

The operation of blowing is often referred to as winnowing or fanning. In its simplest form the uncleaned seed is thrown into the air on a windy day. The components separate out and the desired ones are retained. Indoors the airstream from a fan can be utilised. Hand winnowing has been used successfully in Thailand to separate full seeds from empty seeds of *Pinus kesiya.*

The initial separation was followed by a second winnowing of the discarded fraction. Cutting tests showed that the original full seed percentage of 82 % was raised to 98 % in the improved fraction, while the discarded fraction (approximately 10 % of the total volume) contained only 18 % full seed or about 2 % of the total full seed in the seed lot. One operator took 8 minutes to winnow 1 kg. Loss of the very smallest and lightest full seed in the separation process is usually not serious since these seeds are likely to germinate slowly and to be low in vigour.

Simple winnowing machines can be easily constructed and Yim illustrates a machine made entirely of wooden components which is used in Korea. Laboratory blowers can be classified as either the pneumatic type—a fan is situated near the air intake and *pushes* air through the system, or the aspirator type—a fan is situated near the air outlet and *pulls* air through the system, creating a partial vacuum. Small laboratory cleaners such as the 'Brabant' cleaner, and the 'Kamas' laboratory aspirator are available.

Another seed blower in common use is the South Dakota Blower. The principle of blowing is that a sample of seed, when suspended in an upward stream of air of a certain velocity, will split into a light fraction and a heavy fraction, the light fraction being carried upward and the heavy fraction falling down. The two fractions are caught apart from each other. The heterogeneity of the heavy fraction can be further reduced by subjecting it to a second blowing at a higher wind velocity.

In that way a light fraction, a medium fraction and a heavy fraction are obtained. The blowing apparatus in the South Dakota Blower consists essentially of a centrifugal blower, the outlet of which is connected to the bottom end of a vertical tube of a few centimeters' internal diameter and about 50 cm in length. The sample is held on a fine wire gauze at the bottom of the tube. A built-in valve allows the wind velocity to be set at a rate that has been found optimal for the species.

The lighter particles are blown up and trapped by baffles near the top of the tube, while the heavier are retained at the bottom. A more complex blower developed in Canada is described by Edwards. It consists of four plexiglas tubes of varying diameters which cause differential air velocity and, by choosing the appropriate combination of tubes, it is possible to use the equipment either to separate seed from chaff or full seeds from empty seeds. It works well with large seeds e.g. *Abies amabilis* but is not suitable for very small, light seeds such as *Betula* or *Chamaecyparis*.

In Zimbabwe a home-made cleaner was constructed by fitting a tapered aluminium sleeve with internal baffles onto a constant speed domestic fan. The narrow end of the sleeve projects over a compartmented receptacle; full seeds fall into the nearest compartment, while the lighter impurities and empty seeds are blown into the further compartments. A similar locally made seed winnowing chamber, which uses a stream of air from an electric fan, has been successfully used in the

Solomon Islands to separate debris and empty seeds from dry seed batches of *Swietenia macrophylla* and *Campnosperma brevipetiolata.*

Many seed cleaning machines use a combination of winnowing and screening. A coarse upper screen removes larger material, a lower fine screen stops the seeds and lets through fine matter, and then the seed fraction passes through a transverse or nearly vertical airstream from a fan to remove chaff and empty seeds. The air-screen cleaner is the basic equipment of seed cleaning plants. The size of air-screen cleaners varies from the small two-screen model to a modern precision model, which uses several top and bottom screens and in one operation as many as three air separations.

Flotation

Cleaning by flotation relies on the principle that the density of the seed of a given species is specific both for filled and unfilled seed. There are two basic methods used:

i) *density method* in which liquids with a density or specific gravity between that of the full and empty seed are used. The specific gravity of the liquids used is usually below 1.0 and such that the full seed sinks and the empty seed and light debris float.

ii) *absorption method* in which water is used and, although both full and empty seeds float initially, after some time the full seeds absorb water, become heavier and sink. The time of soaking can vary fro.n a few minutes to several hours. This method is useful where there is a very small difference between the specific gravities of the full and empty seeds. The seed must be re-dried after being separated.

The flotation methods can separate out insect attacked, mechanically damaged, and immature seeds from filled mature seeds. The density method can only be applied if a liquid of suitable density is available which is not harmful to the seed.

Friction Cleaning

Most debris can be removed from the seed by air-screen combinations, but leaf fragments, resin particles and other objects of similar size and density to the seed are difficult to remove. Friction cleaning relies on the principle that any object falling or sliding over a surface encounters a certain friction. The movement of the particle is proportional to its weight and to a coefficient of friction which depends on the nature of the particle's surface and the surface on which it moves.

Separation of seed from debris is made on an inclined cloth or rubber belt on the basis that the angle necessary for the run-off of the seed differs from the angle necessary for run-off of the debris. A continuous upward moving belt removes seeds downwards by gravity and the lighter debris upwards by friction.

Specific Gravity Separation

This method makes use of a combination of weight and surface characteristics of the particles to be separated. It is a method which is finding increasing use in separating and grading tree seeds. The specific gravity (SG) separator employs a flotation principle. A mixture of seeds is fed onto the lower end of a sloping perforated table. Air, forced up through the porous deck surface and the bed of seeds by a fan, stratifies the seeds in layers according to density, with the lightest seeds and particles of inert matter at the top and the heaviest at the bottom.

An oscillating movement of the table causes the seeds to move at different rates across the deck; the lightest seeds float down under gravity and are discharged at the lower end, whilst the heaviest ones are kicked up the slope by contact with the oscillating deck and are discharged at the upper end. Deck coverings of linen, plastic, and wire mesh have been used to distribute the air uniformly beneath the seeds and provide the correct push to the heavier seed in contact with the deck. A closely woven covering such as linen gives the best results for small seeds.

On modern gravity separators, it is possible to control the feed speed, the tilt of the table in two directions, the speed of oscillation and the force of air differentially at various points along the deck. The combination of these different controls makes it possible to adapt the machine to handle a wide variety of species and seed lots. The specific gravity separator will separate particles of the same density but of different size, and particles of the same size but of different densities. It will not separate efficiently particles which differ both in density and size, i.e. separate a larger but less dense particle from a smaller but denser particle.

It has been found practical for cleaning the chaff from some eucalypt seeds and for grading pine seeds. Purity % of uncleaned seed of *E. grandis* after extraction is about 10 %. This was raised to 95 % purity with 95 % germination by SG separator treatment.

Other Cleaning Methods

A number of other methods of seed cleaning have been used experimentally, but are not yet in widespread operational use. They include electronic and electrostatic separators, magnetic separators, electronic colour separators and shaking tables, which separate seeds by the angle at which they rebound when thrown against fixed walls. Seeds of *Ochroma* can be successfully cleared of their floss by placing the uncleaned mass on a wire sieve of 0.3 mm mesh and setting fire to the floss Goor and Barney. The fire flashes across and the seeds drop through the mesh.

Good results are obtained when the seeds are allowed to drop through the mesh into a pan of water. The inflammable oil in the floss burns with intense heat and experience in Honduras has shown that the floss must be spread out thinly to avoid damaging the seed. This method has also been tried to clear *Populus* seeds of floss but may damage more than 50 % of the seeds in this genus. Seeds of *Prosopis* are often embedded in a gummy matrix within the pods. One way of obtaining clean seeds is

(1) To remove one side of the pod mechanically with a knife

(2) To soak the pod contents in a 0.1 normal solution of hydrochloric acid for 24 hours

(3) To wash in water for one hour, then dry in the direct sunlight

(4) To beat or pound the dried mass to separate the clean seeds from the gummy coating.

This method has been successful in India, yielding clean seeds with a germination rate of 65 % in 12 days.

Grading

Within a single species there is a variation of seed dimensions due to environmental influences during the development of the seed and to normal genetic variability. The performance of the seed immediately after germination is related to seed size and, in order to produce a crop of seedlings which will emerge and grow evenly in the nursery, size grading of seeds can be a useful practice. Size grading can also assist mechanical sowing of seeds. The practice should be used with caution in the case of seeds collected from seed orchards having a restricted number of clones.

Since part of the variation in seed size and shape is genetic, grading of seed orchard seed could lead to excessive genetic differentiation and loss of genetic diversity within each of the fractions obtained by grading. The methods of grading seed vary little from those used in the cleaning process. Sieving and screening, cylinders, air blowing, flotation and specific gravity separation can all be used effectively for size grading tree seeds.

Although grading itself is a relatively simple operation, seeds of certain species have to be properly processed before they can be graded, for example the spongy exocarp of teak fruits and the calyx tubes of dipterocarps must be removed in order to get the full benefits of grading. The cleaned stones of *Gmelina arborea* have been graded by using square-mesh screens of 7,9 and 11 mm mesh; germination varied from 84 % for the smallest to 111 % for the largest size class (there are usually 1 to 3 seeds per stone).

Moisture Content Control

After seeds have been cleaned and graded, they are ready for sowing in the nursery. If, however, they are to be put into storage, it is necessary to check their moisture content (MC) and, if necessary, adjust it to the optimum level for storage of the species in question. Adequate facilities for testing MC should be available in the seed processing depot.

For orthodox seeds, which comprise most coniferous and many hardwood seeds, adjustment of MC if needed means further drying. Much less commonly and only in the case of recalcitrant seeds which must be stored at a high MC it may be necessary to moisten the seeds in order to raise MC to the optimum for storage.

For example promising results have been achieved for Acer pseudoplatanus by soaking the seeds in water for two or three days and immediately afterwards freezing and storing them at about -7°C in plastic sacks. Other genera, such as Quercus and Castanea, which have been lightly dried under cover in order to loosen their husks or involucral attachments, may benefit from soaking to restore MC to the optimum level, before being placed in moist, cool storage.

Seeds, like cones and fruits, are hygroscopic materials and, when detached from the parent tree, they lose or gain moisture to or from the surrounding atmosphere until their moisture content (MC) reaches a point of equilibrium with the humidity and temperature of the surrounding air. This is known as the equilibrium moisture content (EMC). Once it has been reached, it will be maintained as long as the humidity and temperature of the air remain constant; if they change, the seeds will again lose or gain moisture until a new EMC is reached.

Wood is another good example of a hygroscopic material and behaves in a similar way. "Wet" seed surrounded by "dry" air will lose moisture and therefore weight, while "dry" seed surrounded by "moist" air will gain them. In order to devise the most suitable methods of drying and storing seed, it is necessary to be able to quantify the moistness of both air and seed.

Air humidity: Moisture in the atmosphere is in the form of water vapour, but air can hold only a limited quantity of water vapour. If this is exceeded, the air is said to be saturated and the excess moisture condenses as dew. The exact weight of water vapour (WV) which the air can hold at saturation depends on the temperature, as shown in the following table:

Temp. °C	-10	0	10	20	30	40	50	60
Weight of WV at saturation (g WV per kg of dry air)	1.6	3.8	7.6	15	27	49	87	152
Density of dry air (kg per m^3) at a pressure of 760 mm	1.34	1.29	1.25	1.20	1.16	1.13	1.09	1.06
Weight of WV at saturation (g WV per m^3 of dry air)	2.1	4.9	9.5	18	31	55	95	161

Most of the time the content of water vapour in the air is less than that at saturation. Relative humidity (RH) is defined as the ratio (usually expressed as a percentage) of the quantity of water vapour actually present in the atmosphere to the quantity which would saturate it at the same temperature; alternatively, as the actual vapour pressure in the air as a percentage of the saturation vapour pressure at the same temperature. For the seedsman relative humidity is the most important measure of atmospheric humidity because the equilibrium moisture content of seeds is most closely correlated with it.

For example the MC of seeds will be very nearly the same when in equilibrium with air at an RH of 50 %, whether the air temperature is at 10°C (absolute humidity or weight of water vapour present = 7.6 / 2 = 3.8 g/kg dry air) or at 50°C (absolute humidity or weight of water vapour present = 87 / 2 = 43.5 g/kg dry air). Though the absolute humidity of one is over ten times that of the other, the relative humidities are the same and it is relative humidity which has the greatest effect on the EMC of seeds.

The importance of RH to the EMC of seeds and the dramatic effect of temperature changes on RH explain the importance of heat in the drying of many seeds.

Moisture content of seeds: The amount of moisture in seeds is usually expressed as a percentage of their weight. Moisture content can be expressed in two ways (a) The weight of water expressed as a percentage of the initial "wet-weight" or "fresh-weight" of the seeds (= dry matter + water) or (b) The weight of water expressed as a percentage of the final oven-dry weight of the seeds (= dry matter only).

One of the greatest difficulties in the understanding and application of published results on moisture content derives from the fact that, in the past, both the "wet-weight" and "dry-weight" methods have been used, often with no indication as to which of them were applied in a particular case. According to the ISTA Regulations *seed moisture content should always be expressed on a wet-weight basis*.

Because of the limited amount of water vapour which it takes to saturate air, a relatively small quantity of seed can hold as much moisture as a great deal of air. One litre of seed dried from 50 % to 9 % MC (wet weight) at 30°C would lose about 450 gms of moisture to the surrounding atmosphere, enough to change the RH of about 15 m3 (or 15,000 times its own volume) of air from 0 to 100 %.

In the case of sun-drying, the atmosphere is so vast that it can absorb this moisture without difficulty, but in an enclosed building the ambient air can quickly become saturated. This explains why so much emphasis is put on adequate ventilation in kiln drying, in order to ensure that moist air is removed as it approaches saturation and is replaced by fresh, dry air. The same feature is an advantage when storing dry seeds in sealed containers.

Provided that the seeds are correctly dried and the containers properly sealed, a relatively small volume of seed will come into equilibrium with a much larger volume of enclosed moist air without increasing its own MC significantly. If one litre of seed of oven-dry specific gravity 0.5, dried to 9 % MC (wet weight), were to be enclosed in a 10 litre sealed container with 9 litres of moist air at 100 % RH at 20°C, the total moisture content of the air would be only:

$$\frac{9 \times 18}{1000} = 0.16\text{g}$$

Even if the seeds were to absorb all this moisture, it would only raise their MC from 50 to 50.16 g or MC % from 9.09 to 9.12 %. The common prescription to fill sealed containers as full as possible with seed is a sound one, but it is based on the deleterious effects on many species of the oxygen in the enclosed atmosphere, not of the water vapour.

Other factors affecting EMC. Although relative humidity is the most important single factor affecting the equilibrium moisture content of seeds, it is not the only one.

1) *Temperature*: As already explained, temperature has a large indirect effect on EMC because, if absolute humidity is kept constant, relative humidity is

directly related to temperature. It has an additional effect because EMC varies slightly with temperature even when relative humidity remains constant. The effect varies with species but very few data on forest trees have been published. An example quoted for an agricultural crop, sorghum, by Justice and Bass shows that at 50 % RH the EMC varies from 12 % at 49°C to 14 % at -1°C. The difference is slightly greater in some other crops, but in all cases EMC *decreases* with increasing temperature and constant RH (even though absolute air humidity *increases* with temperature at the same RH).

2) *Absorption and desorption.* For any species there is a difference of 1 - 2 % in the EMC according to whether a moist seed is losing moisture to a drier atmosphere (desorption) or a dry seed is gaining moisture from a moister atmosphere (absorption). The EMC is always higher on desorption and it is the desorption curve which is important in the common situation of drying orthodox seeds from a higher to a lower MC for storage.

3) *Variation in EMC according to species.* The MC of seeds in equilibrium with a given RH and temperature varies with species. The EMC for each species must be determined by trial. An important component in interspecific variation is the percentage of oil content in the seeds. Seeds which store most of their food reserves as proteins or starch have a higher EMC at a given RH than seeds which store food as fats and oils, because the former are relatively hydrophilic, the latter hydrophobic. Among agricultural seeds wheat, with a low oil content of 2% and an EMC at 45%

RH and 25°C of 10.4%, may be compared with *Brassica oleracea*, with a high oil content of 35 % and an EMC under the same conditions of 6.0 %.

Drying of Orthodox Seeds

For medium- or long-term storage of many species, a moisture content of 4–8 % is recommended. This is considerably less than the MC of freshly collected seeds. Reduction of MC can be achieved in most species by placing the seeds in an ambient atmosphere of relative humidity (RH) 15–20% for a period sufficiently long to allow the seeds to reach an MC in equilibrium with the RH.

The effectiveness of air-drying depends on local climatic conditions. Reduction to an MC of 12–18 % is frequently possible, provided that attention is paid to adequate aeration of the seed. Reduction to less than 8 % is impossible in most temperate situations and in some areas of the wet tropics, because average RH remains too high. Thus RH in moist tropical West Africa is commonly over 80 % in the wet season and over 70 % in the "dry" season.

An MC of less than 8 % is unlikely to be achieved in these conditions. In areas where insolation is high, many species can be successfully dried to 6–8 % MC by exposing them to direct sunlight, because the seeds and surrounding microclimate heat up and RH is consequently reduced. Also, for a constant RH, the equilibrium MC of seeds decreases with increasing temperature. Care must be taken to ensure that the seeds are as dry as possible prior to exposure and that they are moved frequently. In Honduras this method works well with *Pinus* spp. but is not recommended for *Cordia* because seed of this genus dries too quickly and can reach 4 % MC, with damage to the tissues.

Artificial drying can be accomplished in two ways. One method is to raise the temperature of the air which, provided that no additional water vapour is introduced from outside the system, automatically reduces the RH. The other is to remove moisture from the air without changing the temperature, which also reduces RH. He quotes the example of air at 5°C and 90 % RH which is heated to 35°C. Its RH is thus reduced to 15 % and it can be blown through the seed by forced ventilation until the latter reaches equilibrium MC.

On the other hand, air at 30°C and 90 % RH, typical of moist tropical areas in the rainy season, would still have an RH of 40 % even if heated to 45°C. High temperatures can be extremely injurious to seeds, especially those which have a high MC. Generally, drying temperatures should not exceed 40°C and recent trends have been towards lower temperatures and greater air-flows to ensure safety in drying. Barner and CATIE recommend a temperature not exceeding 30°C in the early stages.

One possibility is to dry seed in two stages, the first to about 11 % MC using a temperature below 40°C, the second to about 5 % using a temperature of 60°C. Provided that the MC is reduced to about 11 % in the first stage, a second stage temperature of 60°C is considered safe for most agricultural species. On the other hand, there is some evidence that a high drying temperature, which does not affect immediate germination, may still affect subsequent longevity. For long-term storage for genetic resources purposes, a combination of low RH and low temperature (15 % RH and 15° C) is recommended.

Where climatic conditions make it impossible to achieve a low enough RH by heating the air, it is necessary to reduce it by removing water vapour without raising the temperature. This can be done by either: (a) refrigerating the air to below the dew point, condensing the water vapour on the cooling coils and then reheating the air to 35°C or (b) blowing air through a chemical desiccant, which removes the water vapour, and then through the seed. Various desiccants are available e.g. silica gel, CaO, H2So4, Lithium chloride or anhydrous CaCl2 but the most indestructible and easiest to reuse is silica gel.

A good example of one drying technique is that furnished by the seed bank of the Regional Genetic Resources Project at Turrialba, Costa Rica. Because of the permanently high air humidity and high day temperatures there, hot air drying is not possible without damage to the seeds. Therefore a silica-gel type dryer, which maintains an RH of less than 15 % at 25°C, is used, situated outside the drying room and ducted into it.

Small lots of seed that have first been air-dried to an MC below 20 % may be placed in a sealed container with an equal quantity of silica gel that has been freshly dried at 175°C and cooled. The silica gel, seed and enclosed air will come to an equilibrium suitable for storage. The period of time needed for forced-draught air circulation to dry seeds to equilibrium moisture content is partly dependent on the accessibility of individual seeds to the air current.

Seeds should be spread in thin layers on trays, with space for air circulation between the trays. The MC of coniferous seeds, which have been extracted from the cones during kiln-drying, may already be close to that recommended for storage. But moisture may be reabsorbed during cleaning and dewinging, and in some species it may be added deliberately to reduce damage during these operations. Therefore MC must be tested and, if necessary, further reduced immediately before storage. At the same time, if cleaning and dewinging are carried out in warm, well-ventilated rooms, this will decrease the amount of drying needed subsequently.

Mixing before Storage

When a large seed lot is to be stored in several containers, it is desirable that seed homogeneity between containers be maintained as far as possible, so that each container is equally representative of the seed lot as a whole. If seeds have been size-graded, mixing is carried out on the ultimate seed lots separated by the grading operation. Thus, a given seed collection may be graded into a "large seed" fraction and a "small seed" fraction. Each fraction becomes a separate seed lot with its own seed lot number and separate mixing within each of these two seed lots may be done before storage to ensure homogeneity between containers.

3

Seed Pre-treatment

Seed Dormancy

A mature, live seed, which will not germinate promptly when placed under suitable conditions, is said to be dormant. The mechanisms that prevent germination vary widely and it is not necessary to understand them to grow trees from seed successfully. What is important for us to understand is how we can get these dormant seeds to germinate.

Relatively simple and effective methods exist to overcome the dormancy of our tree and shrub seeds. The most common requirements are exposure to periods of warmth and/or cold. The simplest method of providing these conditions is to place the seeds outdoors, mixed with a moisture retentive and aerated medium (usually a sand/compost mixture).

Seed Pre-treatment

In order to be successfully treated seeds must be moist. Seeds that are placed into treatment immediately following extraction will require no further processing. Seeds that have been stored for any length of time will require soaking in water for 24-48 hours. Soaking dried seeds performs two functions. Firstly, the seeds take up water and, secondly, some of the germination inhibitors in the seed coat are removed.

Stratification Method

Seeds of most species should be mixed with a moisture retentive and aerated medium in a 50:50 ratio of seeds and mixture by volume. The ideal mixture is 50% coarse, sieved peat-alternative and 50% coarse sand. The mixture should be moist but not waterlogged.

Water is added to the dry ingredients, thoroughly mixed and tested by squeezing a handful. When only one drop of liquid can be squeezed out between the fingers then the mixture is ready. This ensures that the seeds remain moist throughout the treatment period, but the exchange of gases (oxygen and carbon dioxide) is not inhibited.

The seeds and/or mixture are placed in a suitable container. Ordinary plant pots (with a layer of sand above and below the seed/compost mixture) are fine. Seeds of alder and birch can be stratified 'naked' (i.e. without the medium). The seeds are placed in cotton bags and soaked in cold water for 48 hours. After a superficial drying, they are placed in polythene bags, which are loosely tied and placed into the fridge for four to eight weeks.

Hot-water Treatment

For broom and gorse seeds a slightly different approach is needed. The procedure is not complicated, but a little unusual. The seeds will have been collected and stored dry. Just before sowing, the seeds are placed in a heatproof container and two or three times their volume of boiling water is carefully poured over them. They are left in the cooling water for another 24 hours. They are then ready for sowing.

Pretreatment to terminate dormancy and speed up germination is thus one important type of pretreatment. Another comprises various forms of seed dressing, pelleting or coating, designed to give the seed protection against pests, diseases or adverse conditions, to improve the uniformity of seeds or to render them more visible to the nurseryman.

Types of Dormancy

Dormancy may be of several different types and sometimes more than one type occurs in the same seed. The simplest grouping distinguishes between

(1) Exogenous or seedcoat/pericarp dormancy
(2) Endogenous or embryo dormancy and
(3) Combined dormancy, in which both seedcoat and embryo dormancy occur at the same time.

In most species of the moist tropical forests dormancy is not a problem. Conditions of temperature, moisture and oxygen are almost invariably suitable for germination immediately on dispersal, so most species germinate within a matter of days or weeks and would not benefit from dormancy. In the drier tropics, on the other hand, seedcoat dormancy is common and some form of pretreatment is necessary to obtain rapid and uniform germination.

Both physical dormancy, caused by hard seedcoats or pericarps with cutinised layers which are impermeable to water, and chemical dormancy, caused by inhibiting chemicals present in the seed covering, occur; it is likely that they sometimes occur simultaneously in the same seed. But it is often difficult to differentiate between them, since a treatment which softens the seedcoat, e.g. hot water, will automatically leach out inhibitors at the same time.

The covering of seeds which exhibit mechanical dormancy is thick, tough and resistant to embryo growth but, unlike the "hard seeds" characteristic of physical dormancy, it is permeable to liquids. The severe treatments, such as boiling water or acid, which can be effective in overcoming physical dormancy, would penetrate the covers of the mechanically dormant seeds and kill the embryos. Mechanical dormancy occurs in a number of temperate genera e.g. *Crataegus, Carpinus, Elaeagnus*; it is not known whether it occurs in the tropics.

In temperate forestry, methods of overcoming the various types of endogenous dormancy have been investigated for a long period. It is doubtful whether any of these methods will have application in the lowland tropics, but some of them may be applicable to species planted in the sub-tropics and tropical highlands; for example germination of *Pinus taeda*, an important species for afforestation in some tropical highlands and in the sub-tropics is known to benefit from cold moist pretreatment. In general, however, the physical and chemical types of exogenous dormancy are likely to be of most concern to tropical foresters.

Seedcoat Dormancy

Seeds of some species have hard, cutinised seedcoats that completely prevent the imbibition of water and sometimes also the exchange of gases. Without imbibition and gas exchange renewal of embryo growth and germination are impossible. Physical seedcoat dormancy of this kind occurs most frequently in species adapted to alternating dry and wet seasons, including several leguminous genera such as *Acacia, Prosopis, Ceratonia, Robinia, Albizzia, Cassia*. In some species, e.g. *Tectona grandis* and *Pterocarpus angolensis*, the hard layer is formed by the pericarp or fruit.

The unit of dispersal, storage and sowing in *Tectona* is the fruit, often referred to loosely as the "seed", but for practical purposes the origin of the impermeable layer does not affect the choice of pretreatment. Pretreatment to overcome physical seedcoat dormancy are designed to soften, puncture, wear away or split the seedcoat in order to render it permeable, without damaging the embryo and endosperm within. They include physical and biological methods, dry heating and soaking in water or chemical solutions.

Any treatment which destroys or reduces seedcoat impermeability is commonly known as scarification. Destruction of impermeability at a single point in the seedcoat is normally sufficient to allow imbibition and gas exchange. Chemical seedcoat dormancy, caused by the presence of chemicals located in the seedcoat but inhibiting germination of the embryo, can usually be overcome by some form of liquid treatment which leaches out the chemicals.

Physical Methods

One of the simplest and most direct physical methods is to cut, drill or file a small hole in the coat of each seed before sowing. This method has been found appropriate in the Philippines for large leguminous seeds in the genera *Afzelia, Albizzia, Intsia* and *Sindora* and in Honduras for *Acacia, Prosopis, Enterolobium* and other legumes. *Intsia* seed is nicked at each end and a third time in the area of the hilum and micropyle; the last area is the most important. In Tanzania the hard and bristly pericarp of *Pterocarpus angolensis* is chipped at one edge by knife or cracked by beating with clubs.

In the Philippines the hard seedcoat of *Eusideroxylon* is cracked by hammer. In *Calophyllum* complete removal of the seedcoat has produced better germination than nicking. Sandpaper may also be used to reduce seedcoat thickness by abrasion. In trials in Pakistan sandpaper proved the most effective treatment in increasing and accelerating germination in a number of hard-coated species. As an example *Leucaena* seeds had zero germination in the untreated control and after 24 hours' soaking in cold water. This was increased to 42 % in 26 days by 1 minute soaking in concentrated H_2SO_4, to 60 % in 13 days by 2 minutes in boiling water, and to 100 % in 3 days by sandpaper treatment.

In laboratory trials in Sweden, sandpaper scarification followed by a three hour cold water soak, was the most effective treatment for *Acacia farnesiana*, yielding 88 % germination in 7 days and 100 % in 21 days, compared with 63 %, 23 % and 3 % in 21 days, respectively for soaking in concentrated sulphuric acid, absolute alcohol and hot water. In six other hard-seeded species from Iraq tested at the same time, it was less effective than one or other of the soaking treatments. Manual treatment of individual seeds is slow but, with experienced workers, is safe and effective. It is best suited to the larger and more refractory seeds e.g. *Delonix regia*. The effect of physical scarification may be reinforced by soaking in cold water before sowing.

For treatment of large quantities of seed, mechanical scarification is more suitable than the manual method. Seed may be tumbled or churned in a concrete mixer with sharp gravel or sand, or in a special drum lined with abrasive material such as sandpaper, cement or crushed glass or incorporating rotating abrasive disks. If gravel or sand is used, it should be sieved to ensure that it can be readily separated from the seed by use of an appropriate mesh screen.

The method is not suitable for seeds with abundant resin or pulp which may clog the machine. Care must be taken to avoid overtreatment causing damage that can reduce or destroy germinability of the seeds. Examination of the seedcoat surface, if necessary by hand lens, or testing the capacity of the seed to take up water, demonstrated by swelling, can be used to estimate effectiveness of scarification.

In India mechanical scarification is used successfully to overcome seedcoat dormancy in *Acacia catechu*, *A. nilotica* subsp. *indica*, *Albizzia falcataria*, *A. lebbek*, *Cassia fistula*, *C. javanica*, *C. nodosa*, *Delonix regia*, *Dichrostachys cinerea*, *Santalum album*, *Terminalia arjuna* and *T. tomentosa*. Horizontal shaking of seeds of *Acacia farnesiana*, bottled with crushed glass, for five hours gave a germination of 20.7 % plus additional imbibed seeds 10.7 % in 43 days, compared with 0.7 % + 0.7 % for the control in the same period.

Soaking

A number of treatments involve soaking seeds in water or other liquids. These wet treatments may combine the effects of softening hard seedcoats and leaching out chemical inhibitors. Some seeds which have little resistance to germination may respond well to soaking for 24 hours in water at ambient temperature. This may be the result of a more rapid imbibition than can be achieved in a moistened nursery bed. It is recommended for some species after manual, mechanical or acid scarification. 48 hours' soaking of *Pinus caribaea* seed in water at room temperature gave more uniform germination than in the untreated seed.

In India soaking in water, for periods varying from two to 48 hours according to species, accelerates germination in *Acacia mearnsii*, *A. melanoxylon*, *A. nilotica* subsp. *kraussiana*, *Adenanthera microsperma*, *Albizzia amara*, *A. procera*, *Grevillea robusta* and *Trewia nudiflora*. A more effective treatment, especially in hot climates, is alternate wetting and drying of seed. This has frequently been used for *Tectona* fruits. In Thailand, after scarification, alternate soaking and drying was applied, 4 times soaking and 3 times drying, each for 30 - 45 minutes.

Elsewhere, if preliminary scarification is not applied, common practice is to spread the "seed" out on a hard surface in the sun in a layer about 5 cm deep and to soak it thoroughly; then to turn it over from time to time, allowing it to dry out and bake in the sun for a day or two. This process of alternate soaking, drying and baking is repeated a number of times, usually between five and ten cycles, until signs of germination appear. Each cycle of wetting and drying may be one day for soaking and 3 - 5 days for drying and baking.

As soon as germination starts, the "seed" should be sown in the nursery. In Tanzania the preliminary soaking is for 72 hours in full sacks in a stream or in large drums. It is then sown on the surface at the rate of 5 kg/m^2 and, after two days in the sun, is covered with about 2 - 5 cm depth of soil and watered daily.

Tectona grandis is one tropical species in which there is evidence that chemical dormancy caused by inhibitors in the pericarp may be more important than physical dormancy. Fairlamb and Davidson found that an aqueous extract, obtained by soaking *T. grandis* fruits for 4 days and used to moisten filter paper, inhibited germination of cress seeds. Germination was 11 % in 144 hours in the extract, compared with 76 % in tank water and 96 % in distilled water. On the other hand Pattanath could find no evidence for physical dormancy in *T. grandis*. She found that, after soaking the fruits in water for 24 hours, water had soaked right through to the locules where the seeds were attached. There is great variation in the degree of dormancy between provenances of teak.

Some require no pretreatment, some require alternate wetting and drying as described above, others respond to soaking for four hours in Sach's nutrient solution, which may indicate a nutrient imbalance in the seeds. Chemical dormancy apparently occurs also in *Terminalia ivorensis*. Alternate daily soaking and drying of the fruits of this species for seven days was effective in producing 50 - 70 % germination, whereas germination of untreated seed is usually poor.

In a later experiment the same author found that an aqueous extract from *Terminalia ivorensis* fruits had a slight inhibitory effect on germination of lettuce seeds but a more serious effect in the number of malformations in the germinated seedlings. He identified Coumarin as one of the more important inhibitants involved. Soaking in running water for one to two weeks has been used successfully to remove inhibitors in *Atriplex* spp.

Hot water treatment has given good results with a number of leguminous seeds. The seeds are usually placed into boiling water which is immediately removed from the heat source and left to cool gradually, the seeds remaining in the water for about 12 hours. They imbibe and swell as the water cools. The proper relationship of the volume of water to volume of seeds can be determined by experiment. It may vary considerably according to species, ratios of 2 – 3 times, 4 – 5 times and 5 -10 times as much water as seed have been suggested.

Some species respond better to an initial temperature well below boiling e.g. *Albizzia falcataria*. An initial temperature of 90°C, cooled to ambient 20°C, has given good results with *Parkinsonia aculeata* and *Ziziphus spina-christi*. Of various initial water temperatures and periods of soaking and cooling tested for *Leucaena leucocephala* in the Philippines, one minute soaking from an initial temperature of 80°C gave the best result 90 % germination. The period of soaking and cooling appeared to have little effect, e.g. an initial 80°C and 6 hours' soak and cool gave 89 % germination.

The initial water temperature had a big effect; germination was only about 30 % after an initial temperature of 100°C and only about 25 % after an initial 40°C. Bowen and Eusebio found that in *Acacia mangium* seeds in Sabah there was a close correlation between the initial temperature of the water and the subsequent germination. Germination increased progressively from 5 % after immersion in water at 30°C to 91 % after immersion at 100°C. Prescribed treatment for this species is now to immerse seeds in 5 x their volume of water at 100°C, stir in the cooling water for 30 seconds, pour off hot water and soak overnight in 20 x their volume at room temperature.

For *Albizzia falcataria* a lower initial temperature of 80°-90°C is recommended. Some species of *Acacia* require more severe treatment. Larsen found that, whereas the standard treatment with boiling water improved germination of *Acacia sieberiana*, an exceptionally resistant species, from 2 % to 10 %, continuous boiling of the seed in water for 60 minutes achieved 60 % germination within two weeks.

Prescriptions for hot water treatment must be applied meticulously if seedcoat dormancy is to be removed without killing the seeds through excessive heating. It is often easier to ensure this at a central research laboratory than in a number of scattered field nurseries. In Sabah it was found possible to redry seeds of *Acacia mangium* and *Albizzia falcataria* after the initial 30 seconds' hot water treatment, package and despatch them to field nurseries and, after 3 days' storage, to give a simple cold water soak before sowing.

Germination was just as good as when seeds were sown immediately after treatment. Drying was done in a current of dry air at 45°C for periods of 6 – 18 hours. Hot water treatment is relatively easy and safe to apply, as well as being effective with some species. It is not well adapted to large lots because of the difficulty in handling and sowing the swollen seeds.

Acid Treatment

The chemical most commonly used to break seedcoat dormancy is concentrated sulphuric acid. For some species it is more effective than hot water treatment. Seed which has been kept for a long period in store may require a longer period in the acid than fresh seed, which could be severely damaged by the same length of treatment. Great care is needed in the handling of sulphuric acid and this method is not suitable for use by unskilled workers.

Materials and equipment required are as follows: commercial grade (specific gravity 1.84, 95 percent pure) sulphuric acid; acid-resistant containers (thick plastic preferred); wire containers and screens for handling, draining, and washing the seeds; an abundant supply of running water; a safe place to drain the dilute acid resulting from rinsing the seeds; and facilities for drying the seeds after rinsing.

Safety precautions are a must! All workmen must understand and obey safety precautions in the use of acid. Seeds, containers, implements, and the acid itself must be handled with great care to avoid injury. Water must not be splashed into the acid, as a violent reaction will occur. All workmen should wear suitable safety clothing, gloves, and goggles or other eye protection.

Toughness of the seedcoat varies between lots and even between individual trees in most species. The optimum period of immersion in acid for each lot may be determined by treating a small sample for different periods and then soaking the lots in water at room temperature for 1 to 5 days (depending on species). The treatment period that yields a high percentage of swollen seeds (by water uptake) without visible injury is the right one.

Oversoaking may pit the seed and even expose the endosperm. Insufficient soaking leaves the seedcoats of most species glossy; coats of correctly treated seeds are dull, but not deeply pitted. If tests reveal only small differences between lots, then all may be lumped together for treatment, unless there are other reasons for keeping them separate (such as seed source distinctions). Large differences between lots should occasion separate treatment. The steps in acid treatment are as follows:

1. Allow seeds to come to air temperature. If they have been removed from cold storage, do not open the container until temperature equilibrium is reached. Moisture will form on cold seeds exposed to warm moist air, and this moisture can react with acid to raise temperature to the danger point.

2. Thoroughly mix seeds to be treated as one lot.

3. Immerse seeds in the acid for the required period, making sure that all are covered. Treatment should

be carried out at 65° to 80°F (18°–27°C), preferably on the upper end of the range. Lower temperatures require longer soaking times than do higher temperatures. Careful stirring will reduce the length of treatment necessary.

4. Remove seeds from the acid and wash them promptly and thoroughly in cool, running water for 5 to 10 minutes to remove all traces of acid. Water should be applied copiously at the start, and the seeds stirred carefully during rinsing.
5. Spread the seeds in a thin layer for drying, unless wet sowing is preferred.

Fifty-pound (20 kg) lots can be treated in screen-wire cylinders (reinforced with heavier wire) that can be lowered into the acid. In this way, most of the acid is retained for reuse. After a short draining period, the seed should be washed. Extra care must be taken in large-scale treatments to avoid excessive temperatures that can damage seeds. There are several advantages of the acid treatment. It is effective for many species and requires little or no special equipment. Cost is reasonable.

Most of the acid can be recovered and reused (unless the acid is poured on a pile of seed). Treated seeds can be held from a week to a month or more before sowing, without appreciable deterioration. Since the process leaves the seeds dry, firm and unswollen, they can be sown with mechanical seeders as well as by hand.

There are also disadvantages. Length of treatment must be carefully determined, and temperature must be carefully controlled, especially in large lots, to prevent serious injury to the seeds. Workmen also face a safety hazard. If a large quantity of one seed lot is to be treated in several batches, it is desirable to standardise treatment as far as possible e.g. by using fresh acid of full strength and exactly the same volume (two volumes of acid to one volume of seed) for every batch. In addition to protective clothing, a concentrated solution of sodium or potassium bicarbonate should be kept handy as an antidote against accidental splashes.

Sulphuric acid treatment has been effective for several temperate and subtropical species, e.g. *Gleditsia triacanthos* (1 hour) and *Ceratonia siliqua* (2 hours). Examples of tropical species which respond well are *Intsia palembanica* (60 minutes of soaking), *Parkia javanica* (15 minutes), *Dialium maingayi* (5 minutes), *Acacia albida* (20 minutes), *Acacia nilotica* (60–80 minutes) and *Acacia senegal* (40 minutes), *Acacia planifrons* (2 hours) and *Prosopis tamarugo* (7 minutes).

Hulled seed of *Pterocarpus angolensis* treated with acid gave 60 % germination between 4 and 19 days after treatment, compared with 16 % between 11 and 37 days in the control. In the Sudan it was found that seeds of *Albizzia lebbek, Cassia fistula* and *Prosopis chilensis* could be stored successfully for a further 3 - 4 months after treatment in H_2SO_4 or hot water. Storage after treatment is also practicable for several species of *Acacia*. Other chemicals have been tested from time to time to overcome seedcoat dormancy, but none have been adopted for largescale use in the same way as hot water or sulphuric acid. They include ethyl and methyl alcohol, xylene, ether, acetone, chloroform, hydrochloric acid, nitric acid and sodium hydroxide.

Biological Methods

In nature animals and micro-organisms are an important factor in the breakdown of seedcoat impermeability. It is difficult to make use of these organisms as a controlled pretreatment of seed, but in a few cases successful results have been obtained.

Seeds of *Acacia senegal* and *Ceratonia siliqua* that have passed through the digestive tracts of goats germinate readily when placed in favourable conditions, because of the action of the strong digestive chemicals. Feeding the pods to penned goats and collecting the seeds from the droppings is a convenient pretreatment for these species. Seeds of some species are said to be regurgitated after partial rumination e.g. *Gmelina arborea*. Troup states that seeds of *Acacia nilotica* are ejected after rumination by sheep and goats but pass right through the digestive tract in cattle.

In either case germination is improved by the digestive action. Termites are an important agent for breaking down seedcoat dormancy in many parts of the tropics. In Thailand teak fruits were spread on the ground in a 5 cm thick layer immediately after collection and covered with cardboard. After about 5 weeks the termites had removed the exocarp and subsequent germination, after alternate wetting and drying, was significantly improved in comparison with fruits sown with intact exocarps.

Termites have been used in a similar way to break down the tough winged and bristly fruit of *Pterocarpus angolensis*. Periodic inspection is essential to ensure that the process is not carried too far. Partial fermentation, which is damaging to many seeds, can be beneficial in overcoming seedcoat dormancy in some. In the southern Sudan fruits of *Tectona grandis* are allowed to lie on the ground through the rainy season for partial fermentation. They are then collected, stratified in a pit with layers of (a) Seed (b) Organic matter (c) Soil and watered daily for 10 days. Satisfactory germination results.

Solar Radiation

Solar radiation is not used alone to promote germination but is an important component of the alternate soaking and drying treatment described above under "Soaking in water". In the seasonal wet and dry tropics, fire is a powerful natural factor in the removal of seedcoat dormancy. A fierce fire will kill the seeds but a light to moderate fire, such as those associated with controlled early burning, will reduce seedcoat impermeability and stimulate germination.

Fire has been used in a number of countries to stimulate germination of *Tectona*. The fruits may be spread thickly on the ground and covered with grass which is burnt off, or they may be lightly scorched by a flame gun. Adjusting the heat of the fire to achieve the maximum effect on the pericarp without damaging the seed embryo requires experience. Similar treatment is used for *Aleurites moluccana* in the Philippines. The nuts are spread evenly on the ground and covered with a 3 cm thick layer of dry *Imperata* grass which is set on fire.

As soon as the grass is burned, the seeds are placed in cold water. The quick change of temperature causes the nuts to crack and they are ready for sowing. An alternative is to sow the nuts at correct spacing with only half their diameter in the soil. A layer of *Imperata* grass is spread over the seed bed and set on fire. After burning, the seedbed is sprinkled with water and the nuts are pushed 2 cm deep into the soil and watered thoroughly.

Mechanical Dormancy

The thick, tough but water-permeable coverings of seeds exhibiting mechanical dormancy prevent embryo growth even when water can be freely imbibed. This mechanical obstruction to germination can be overcome by a period of "warm moist" treatment the length of which varies according to species.

The period of "warm moist" treatment can be shortened in some species by a preliminary treatment in sulphuric acid. This requires more care and expertise than the use of acid for physical dormancy. Seeds or fruits must be thoroughly dried before treatment and the process should be limited to the partial digestion of the outer layers only, leaving the final weakening of the inner layers to be done by the subsequent warm moist treatment. In most cases it is preferable to accept the safer but slower method of warm moist treatment alone. Periods of treatment vary from 2 weeks for some species of *Prunus* to 16 weeks for some species of *Crataegus*.

At the end of the appropriate period seeds which possess only mechanical dormancy are ready for sowing. Many species in this class also have physiological dormancy of the embryo and will need further treatment to remove this. It may be noted that the warm moist treatment which removes mechanical dormancy is identical with the warm moist treatment which removes morphological dormancy (underdeveloped embryos).

Embryo Dormancy

Embryo dormancy is a prominent feature in some temperate genera. For example in the USA Rudolf noted that, of 400 species of woody plants studied, about 60 % required pretreatment to induce prompt germination. Endogenous dormancy occurs in both orthodox and recalcitrant seeds. Its occurrence in the lowland tropics is probably rare (most seeds in moist tropical forest germinate quickly or not at all and those in the dry tropics typically have seedcoat dormancy). It could be of importance in the high altitude tropics and sub-tropics.

Trials of prechilling on seed of the Zambales (Philippines) provenance of *Pinus merkusii* showed no improvement in germination; if anything germination was slightly depressed. Wunder mentions *Cordia africana* as a species suitable for prechilling for several weeks in moist sand at 5°C but does not give any comparative figures of germination from treated and untreated seeds.

In *Eucalyptus* prechilling is effective in promoting germination in only a few species from the cold temperate region e.g. *E. delegatensis* and *E. pauciflora;* in the great majority of species it has no effect at all. Endogenous dormancy includes the cases of embryos which are morphologically underdeveloped at the time of separation from the parent tree and which need a subsequent period for further growth before they can germinate. It also includes the cases of embryos which are morphologically mature at the time of seed dispersal or collection but are physiologically incapable of germination until certain biochemical changes, still little understood, take place.

Morphological Dormancy

Seeds with underdeveloped embryos at the time of dispersal or collection will not germinate until the embryos have had time to mature. In a few species maturation is possible during dry storage e.g. *Gingko biloba*. More commonly, a period of moist warm pretreatment is necessary before the embryos develop sufficiently for germination to take place. The recommended treatment is the same as that for mechanical dormancy i.e.

1. Soak seeds in several times their volume of cold water at approximately 3–5 °C for 48 hours.

2. Drain off the water and mix the seeds with two to four times their volume of a moistened, water-retaining medium such as sand, sand/peat mixture or vermiculite.
3. Store at a warm temperature. A constant 20–25°C or alternating 20° and 30°C is suitable for many species.
4. Open containers weekly, mix seeds and, if surfaces show signs of drying out, remoisten with water spray.

Most species which have underdeveloped embryos also have physiological dormancy, therefore warm moist treatment must be followed by cold moist treatment.

Physiological Dormancy

Far more common than morphological dormancy among temperate species are the cases where seeds are fully developed at dispersal or collection but are inhibited from immediate germination for physiological reasons. The pretreatment most effective in removing this physiological dormancy is that which approximates to the conditions of overwintering in nature i.e. a moist cold treatment, or cold stratification.

Cold stratification will not only overcome physiological dormancy but can also reduce the sensitivity of both dormant and non-dormant seeds to their optimum requirements for light and temperature, resulting in increased rate and uniformity of germination over a variety of conditions. If properly carried out, cold stratification does no damage to non-dormant seeds which are undamaged and have not deteriorated from excessive physiological ageing. It can therefore be safely used when different degrees of dormancy are expected in the same seed lot.

Stratification (sensu stricto) refers to the method of placing seeds in layers alternating with layers of a moisture retaining medium, such as sand, peat or vermiculite, and keeping them at a cool temperature for a period, which is commonly between 20 and 60 days but varies considerably from species to species. The combination of high moisture and low temperature appears to trigger off biochemical changes which transform complex food substances into simpler forms utilised by the embryo when it renews growth at germination.

The use of the word "stratification" has recently been extended to include all forms of cold, moist treatment whether or not the seeds are placed in layers. The three main requirements for success in both stratification and moist prechilling are a renewable source of moisture for the seeds, low temperature and adequate aeration.

Only imbibed seeds will benefit fully from cold, moist treatment, while good aeration is needed to supply oxygen for respiration and to dissipate heat and CO_2. Low temperature not only favours biochemical changes in the seed but reduces micro-organism activity and the risk of overheating and premature germination of the after-ripened seeds. In temperate climates low enough temperatures to accomplish prechilling are experienced by overwinter stratification in outdoor pits. The method used in the UK is described by Aldhous. Its main features are:

1. The pit should be located in a cool, shady and well-drained site. The bottom 10 cm should be filled with either sand or gravel for drainage. Pit contents should remain moist during stratification, but should never become waterlogged.

2. 60–80 cm is convenient for depth and width, the length can be adjusted to the volume of seed to be stratified.
3. The pit should be lined at the bottom and sides with a frame carrying 6 mm mesh wire netting as a protection against mice. After the pit is filled, it should be covered by a lid of the same mesh netting.
4. The seed to be stratified should be mixed with 4 times its weight of sand and the pit filled with the seed/sand mixture to within 15 cm of the surface. The top 15 cm should be filled with pure sand.
5. The start of stratification should be timed in relation to the anticipated sowing date and the optimum period of stratification expected for the species. Seeds should be inspected periodically, starting a few weeks before the sowing date. They should be sown when most seeds are beginning to split and the tips of the radicles are visible but before the radicles have elongated. Delayed sowing leads to broken or injured radicles; hardwoods can sustain such damage without too serious ill effects, but it can cause severe losses in conifers.

Where cold-room facilities are available, stratification may be carried out indoors, where closer control of moisture and temperature can be achieved than with the pit method. A temperature of +1° to +5°C is commonly recommended. In the USA one method in common use is to place 10- to 25-pound (4.5 - 12 kg) lots of seeds in loosely woven bags which are flattened into disks no more than 3 inches (7.5 cm) thick and alternated with layers of moist medium. Boxes, trays, cans or drums make suitable containers but must have perforated bottoms to facilitate drainage and aeration.

Soaking seeds overnight in water at room temperature is common practice with southern pines and hardwoods. Containers should be covered loosely to prevent the seed and medium drying unevenly. The seeds should be inspected periodically to prevent heating, poor aeration and excessive drying and to detect the first stages of germination. Seeds should be sown soon after removal from stratification. Stratified but ungerminated seeds of some genera, e.g. *Prunus,* may undergo a secondary dormancy if subjected to extreme drying or to temperatures over 20°C. A new cold stratification is then needed to overcome the secondary dormancy.

Prechilling Methods

Similar results to stratification in layers may be obtained with many species by storing the seeds moist in polythene bags. As with indoor stratification, seeds should be soaked in several times their volume of water before prechilling, a 48 hour soak at 3–5 °C is suitable for many temperate broadleaved species. After soaking, the water is drained off and the moist seeds are then prechilled at 3–5 °C for the period appropriate to each species.

Prechilling may be "naked" i.e. without any medium, or the seed may be mixed with 2–4 times its volume of a medium such as moist sand, moist peat or a mixture of the two Polythene bags of about 100 micron thickness make suitable containers since they are moisture proof but somewhat permeable to oxygen. They should be lightly tied and opened weekly when the seed should be mixed and, if necessary, remoistened. A smell of alcohol on opening a bag indicates that anaerobic respiration is taking place because of inadequate oxygen; in this case frequency of opening and mixing should be increased.

Naked prechilling has the advantage that it is easier to check the condition of the seeds throughout the prechilling operation and there is no need to separate seeds from medium at the end of treatment. On the other hand, there is evidence that germination in some species benefits from use of a medium.

Periods of prechilling vary considerably from species to species and, to some extent, from seed lot to seed lot within a species. For *Abies* a period of 3 weeks at 3–5 °C has proved satisfactory. The same temperature and period is effective for most of the cold-temperate eucalypts, but some provenances of *E. delegatensis* need 4–8 weeks for fast, uniform germination. For *Nothofagus obliqua* and *N. procera* naked prechilling at 3° - 5°C for six weeks, and surface drying before sowing, gave high germination (usually over 80% in 28 days) under nursery conditions. But, treatment with gibberellic acid was a reliable and simpler alternative.

In species having deep physiological dormancy the prechilling period may be as long as 20 weeks e.g. *Liriodendron tulipifera*. For *Fagus sylvatica* in Poland, after storage for several years at -5°C and 10% MC, the following pretreatments are recommended:

1. Allow to defrost.
2. Moisten by sprinkling of water and thorough mixing of wetted nuts twice a day for 6 days, at 3°C, until MC rises to 31%.
3. Leave wet nuts in unsealed containers without any storage medium at 3°C for a period equal to two weeks longer than the minimum required to produce 10 % germination in a sample within two weeks of its transfer to a moist germinating medium. Recurrent sampling and germination

testing is needed to estimate this period, which may vary considerably from seed lot to seed lot. MC of 31% to be maintained meanwhile by periodic weighing of containers and remoistening of nuts to restore loss of weight.

4. Sow in moist germinating medium at 3°C and leave for two weeks. This should initiate germination of radicles.
5. Transfer to temperature of 20°C to promote hypocotyl and epicotyl elongation and seedling emergence, which are inhibited in this species at 3°C.

For large-scale operational sowings, the ideal experimental conditions described above can be simulated to some extent in the nursery by timing the spring sowing so that the nuts experience an adequate period of cold temperature first, followed by the higher temperatures of late spring and early summer. There is some evidence that steps (2) and (3) of the pretreatment described above can be carried out before, as well as after, storage. The advantage is that the nuts are ready for sowing as soon as they come out of store, without the need for a subsequent cold, moist treatment covering several weeks. Nuts pretreated before storage have been successfully stored for 15 months in France.

Chemical Treatment

A wide range of chemicals have been tested experimentally in an attempt to overcome internal dormancy. They include gibberellic acid, citric acid, hydrogen peroxide and a number of other compounds. Some have given a degree of improvement, e.g. Bachelard found that the germination of dormant seeds of *Eucalyptus delegatensis*, *E. fastigata* and *E. regnans* could be improved by treatment with gibberellic acid (GA). 24 hours' immersion in either GA 3 or GA 4/7 of *Nothofagus obliqua* has given rapid and complete germination in 14 days, although this normally dormant species otherwise requires 28 - 42 days' stratification.

The strength of the gibberellic acid had only a small effect, 200 ppm giving 100 % germination in 8 days, 50 ppm the same germination in 12 days. The best stratification treatment (42 days at 3 - 5°C) yielded 70 % in 14 days and 88 % in 28 days and the control (24 hours' soaking in distilled water and no prechilling) only 20 % in 28 days. The saving in time effected by the GA treatment (1 + 12 days compared with 42 + 28 days) is considerable.

Excellent germination was obtained throughout the temperature range 15–30°C, whereas GA 3 required temperatures over 21°C for comparable results. These successes, however, are the exception. In general, for cheapness and reliability chemical treatments cannot compete with stratification or moist prechilling and they are unlikely to play a major role in normal nursery practice in the foreseeable future.

Other Treatments for Endogenous Dormancy

X-rays, gamma rays, light rays in the red region of the spectrum and high frequency sound waves have all been used experimentally to try to overcome dormancy and stimulate germination. Improvement has been reported in some species including *Tectona*, but it has proved difficult to achieve consistent results and the treatments may induce chromosome damage and other abnormalities. Lynn concluded that there was far more evidence of unfavourable effects from irradiation of seed than beneficial effects. None of these methods are suitable for practical application at the present time.

Treatments for Overcome Double Dormancy

Some species combine more than one form of dormancy at the same time. Pretreatment to break one type of dormancy alone will be largely ineffective, unless it is followed by a second pre-treatment to overcome the other type. Physical dormancy of the seedcoat may be combined with physiological dormancy of the embryo. In this case seedcoats should be treated first e.g. by scarification and cold moist prechilling applied afterwards to overcome embryo dormancy. *Cercis canadensis* provides a good example; single treatments gave less than 10 % germination in all cases, while successive applications of seedcoat and internal dormancy treatments gave rapid germination of 45 % (mechanical scarification) or 65 % (acid scarification).

In a few cases treatment of the seed covering after stratification has proved effective. Recent trials with *Fraxinus pennsylvanica* showed that cold stratification at 4°C for 88 days produced 35 % germination in 3 weeks, compared with 2 % in unstratified seed. But removal of the pericarp after stratification further increased germination to 56 %, while the best, but laborious, treatment of stratification followed by removal of embryos from pericarp and endosperm gave 88 %. Treatment to increase permeability of the pericarp in acorns after stratification benefits germination.

Some researchers found that 6 hours of mechanical shaking in distilled water, after 3 months' storage at 0°C and 30 - 40 % MC gave 90 % germination in *Quercus shumardii*, compared with 70 % from 6 hours' simple soaking after storage. The shaking treatment also increased MC over that from the soaking treatment, indicating increased permeability.

Fraxinus excelsior combines morphological dormancy (underdeveloped embryo) with physiological dormancy. Warm moist treatment to remove the morphological dormancy should be followed by cold moist treatment to remove the physiological dormancy.

A number of species, especially in the family Rosaceae, combine mechanical dormancy, caused by a tough, thick pericarp, with physiological dormancy. The same combination of warm moist followed by cold moist pretreatment is effective. In the case of a resistant species, *Crataegus monogyna,* the recommended periods are 4–8 weeks warm, followed by 12–16 weeks cold.

In some species, germination of the radicle takes place readily at warm temperature, but the epicotyl will not start to grow until (a) The radicle has already started to germinate and (b) The seed has been exposed to a period of low temperature. Examples are *Viburnum opulus* and *Carpinus caroliniana*. Here again the same sequence of warm moist treatment at 20–25°C followed by cold moist treatment at 3–5 °C should induce satisfactory germination. Thus the same combination of warm moist plus cold moist treatments applied in that order is capable of removing several different combinations of double dormancy.

Dressing and Pelleting of Seed

In seed dressing or seed pelleting treatments, the surface of the seeds is coated with some inert material, the sticker, to which chemicals of various kinds may be added. Some of the advantages claimed for pelleting and listed by Magini are:

1. The incorporation of fertiliser material in the pellet furnishes the young germinated seedlings with needed nourishment.

2. Plant growth regulators or stimulants may promote rooting or hasten emergence of the seedlings.
3. Fungicides and insecticides are more effective when in direct contact with the seeds.
4. Seeds may be protected against rodents by adding unpalatable, repellent or poisonous substances.
5. Small seeds become larger and heavier and this improves the scatter pattern in aerial seeding.

Pelleting can also be used to incorporate a brightly coloured substance with the seed, which renders it more visible against the soil and thus facilitates uniform sowing rates. Coating with an anti-desiccant material such as alginate can preserve the seed against drying out, until adequate rain has fallen to permit safe germination. Investigations have shown that some fungicides used in seed dressings may have an adverse influence on germination.

The adverse effect was found to increase with temperature and this calls for extra caution in tropical countries. The cost of seed dressings is seldom justified when plants are raised in a nursery. Fertilisers, fungicides or insecticides can be applied more conveniently to the nursery soil than to the seeds, while good nursery hygiene coupled with presence of nursery staff through the day and periodic trapping or poison-baiting should be effective against birds and rodents.

Pelleting is, however, sometimes used to improve the uniformity of seed for precision sowing in the nursery. The main use of pelleting is for direct seeding, including aerial seeding. Protective measures to assist individual seeds after sowing are impracticable and pelleting is the only possible means of achieving some degree of protection.

Most emphasis is now given to seed protection through incorporation of fungicides, insecticides and repellents; inclusion of fertilisers is seldom practised. One example of successful use was in aerial seeding of pines in the southern USA. The coating formulation consisted of endrin and arasan as the protectants, with a latex sticker to act as a bond. The seedling yields in field studies comparing coated with untreated seed were 55 to 1 for *Pinus palustris* and 12 to 1 for *P. taeda*.

Seed coating of eucalypts is practised for aerial seeding of cutover mountain forests in southern Australia, where an area of 8,000 to 12,000 ha annually has been regenerated by this method. Aerial seeding has been used very little in the tropics but some success was obtained in Indonesian trials in Central and Eastern Java on areas dominated by *Imperata* grass, using *Leucaena leucocephala, Calliandra calothyrsus* and *Acacia auriculiformis*.

Latex, methylcellulose and hydrol emulsions are used as stickers. Flaked aluminium powder is sometimes incorporated, because its shining quality is effective in scaring off birds, or to hasten drying and prevent clumping of the treated seed. A variety of chemical protectants is available. Thiram or arasan, endrin and anthroquinone are among those commonly used. Thiram has been found effective against damping off. Red lead was long used as a dressing to protect seeds against birds and rodents but it is now considered to be valueless in this role. It has reduced germination and growth in *Larix* and may sometimes be toxic to the operator in hand-sowing techniques.

A more suitable dye which is equally effective as colouring and which has no apparent adverse effect on subsequent germination is a Waxoline dye such as Lithofar red or the later Waxoline red A. It can also be used as a marker for special certified seed batches in storage. Magini describes one method of coating seeds through use of a small cement mixer. The seed is placed in the mixer (a batch of about 12 kg at a time is suitable) and moistened with a latex sticker solution of one part latex to nine parts of water at the rate of one eighth to one quarter of a litre per kg of seed.

Enough treating powder is then added to dry the sticker, usually in a volume relationship of four parts powder to one part sticker. The thickness of the seed coating is dependent on the amount of sticker in relation to the amount of seed. The total mixing time should not exceed four minutes, since prolonged agitation damages seeds or chips off the pelleted coat.

The procedure described by Aldhous for coating seed with a Waxoline red dye is first to prepare a mixture of 1 part of dye to 19 parts of French chalk. The seed is mixed with a small volume of linseed oil and stirred gently until all the seed appears evenly covered. The freshly oiled seed should be mixed without delay with the French chalk/dye in a closed container until evenly coloured. The seed is then ready for sowing.

In the Philippines tests have been made of Arasan as a rodent-repellent seed coating for aerial seeding of *Leucaena leucocephala*. A sticker solution consisting of 1 part Dow Latex 512 R and 9 parts water was added to the seeds and the mixture stirred for 2 minutes. Arasan 75 wettable powder was then added gradually and the mixture stirred for a further 4 minutes, then spread out to dry for 14 hours. Several different strengths of Arasan were compared with a control. It was found that Arasan did not effect germination adversely, in fact the 7.5 % active ingredient Arasan gave significantly better germination than the control (74 % compared with 64 %).

Results from 5 % and 10 % acid equivalent Arasan were intermediate. Previous trials had demonstrated that Arasan was effective in reducing rodent damage. In Honduras the following mixture has proved suitable for *Pinus oocarpa* and *P. caribaea* direct sown: 60 g Arasan, 20 g 50% Endrin, 5 ml latex and 100 ml water for 1 kg of pure seed. This concentration is considerably lower than that recommended for southern pines in the USA, since it was found that the higher concentrations caused some damage to these tropical species.

Other Types of Pretreatment

One pretreatment designed to separate viable full seed from non-viable full seed has been described by Simak. It involves a form of pregermination of the sound seed and is thus a treatment suitable for application between storage and sowing and not between processing and storage. It is appropriate in cases where seed must be sown after a long period of storage, either because of a long interval between good seed years—a decade or more sometimes in northern conifers—or because of storage for conserving gene resources. Seed stored for such long periods may contain a substantial proportion of full seeds which have lost the power to germinate and the separation of non-viable from viable seeds greatly facilitates subsequent nursery operations. The method involves three stages:

(1) The provision of conditions ideal for initiating the internal seed processes which lead to germination in viable seeds, especially the imbibition of water

(2) Partial redrying of the seeds. The viable seeds retain more absorbed water after redrying than the dead filled seeds and their specific gravity is therefore greater.

(3) Separation, e.g. by flotation, of the non-viable from the viable seeds.

In one seed lot of *Pinus sylvestris* the treatments were

(1) 16 hours' soaking in water, followed by 72 hours' storage in an incubator in a 2 cm thick layer at 15°C and 100% moisture

(2) 12 hours' drying at 15°C, 35% RH and light of 800 lux

(3) Separation into two fractions of "floaters" and "sinkers" by flotation in water.

This method, applied to 10 kg of seed of 67% germination, separated the seeds into one fraction of 7.3 kg "sinkers" with 90 % germination and a second fraction of 2.7 kg "floaters" with 13% germination. The resulting component with the higher germination can be sown at the rate of a single seed per paper pot in the nursery and germination is more rapid because of the pregermination treatment. The method, known as the IDS (Incubation-Drying-Separation) treatment, requires very close control of temperature and moisture conditions if the desired differentiation between live and dead seeds is to be achieved without bringing the pregermination effect on the live seeds to the stage of radicle emergence.

The optimum times of treatment may vary from seed lot to seed lot and it is necessary to carry out a preliminary trial on a small sample of each lot before treating it in bulk. The method is likely to have most application in circumstances where

(1) A substantial proportion of full but non-viable seeds is expected after storage

(2) Close control of temperature and moisture conditions is possible

(3) The viable seed obtained after separation can be sown in the nursery with minimum delay.

It is also possible to separate full but mechanically damaged seeds from full undamaged seeds. One method described by Lestander and Bergsten consists of placing the dry seeds in a drum partially filled with water. Centrifugal pressure is applied to the system by revolving the drum (5000 revolutions per minute has proved an effective speed) for a period sufficient to cause the damaged seeds to absorb enough water to make them sink, while the undamaged seeds absorb very little water and remain floating.

In a seed lot of *Pinus sylvestris* known to have 26% mechanically damaged seeds, it was found that 20% of the seeds had sunk after 1 minute of drum rotation and 25% after 5 minutes. About 98% of the sinkers were damaged and only 2% of the floaters. Germination in the sinkers was about 3% and in the floaters 85% (after 5 minutes' treatment and 21 days' germination test period). The same effect can be obtained by applying pressure to the liquid directly instead of by centrifuging, or by first applying a vacuum and then releasing it. The method is known as the PREVAC (Pressure-Vacuum) method.

This treatment is based on the fact that seeds with undamaged testas absorb water less readily than those with damaged testas, whereas the IDS treatment is based on the fact that live imbibed seed releases moisture less readily on drying than a moistened dead seed.

The PREVAC and IDS methods have been successfully used in combination to improve seed quality in *Pinus caribaea* and *P. oocarpa*. In *P. caribaea* the combined treatments increased germination after 21 days from 75% in the control to 87% (and to 91% after a second incubation period), in *P. oocarpa* from 93% in the control to 99%. Speed of germination after sowing was also improved from 4% after 7 days to 35% in *P. caribaea,* and from 6% to 58% in *P. oocarpa*. The schedules used were those devised for *P. sylvestris* and further improvement in results may be obtainable by adapting these to the tropical pines.

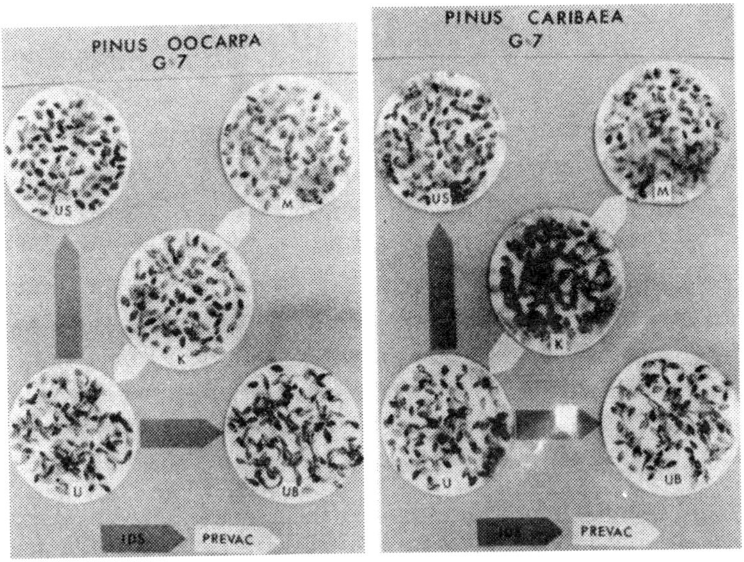

The germinating seeds of IDS/PREVAC-treated Pinus caribaea and P. oocarpa after 7 days.

In the initial experiment about half of the seeds separated as "mechanically damaged" or as "filled dead" actually germinated in *P. oocarpa* and about one third in *P. caribaea;* on the other hand a much higher proportion of these "rejected but germinated" seeds showed signs of abnormal germination or reduced vigour than in the control.

4

Seed Dormancy

Dormancy is the suppression of germination under favourable environmental conditions. An estimated 10 percent of tropical species show dormancy. Several conditions cause dormancy: the presence of rudimentary embryos or physiologically immature embryos, mechanical resistance or impermeable seed coats, endogenous growth regulators inhibiting germination, or inadequate storage; some dormancies are the product of multi-factorial interactions.

Dormancy can be innate or induced. Innate dormancy (primary) prevents the germination of seeds during their development and maturation in the maternal tree and usually some time after dispersal or collection. Dormancy is innate external (primary external) when the seed coat is hard, impermeable to gases or water, or mechanically resistant (e.g., Ilex, Magnolia, Enterolobium cyclocarpum, Samanea saman, Stryphnodendron microstachyum). This phenomenon also applies to seeds enclosed by a hard, woody pericarp or endocarp.

According to Murdoch and Ellis, the hard seeds are innately quiescent and the environment can reinforce the quiescence. When the embryo contains inhibitory substances or it is physiologically immature, the dormancy is innate internal or primary internal (e.g., Juniperus virginiana).

The heritability of dormancy is complex because the distinct parts of the seed are genetically different. Innate dormancy varies with both the genotype and the environment at maturation. Seeds produced in hot, dry summers are less dormant than those produced in cold, humid summers. Those maturing inside green tissues tend to be more sensitive to light than those in which the chlorophyll decreases in the early stages of maturation. The innate dormancy declines before or after dehiscence. This period is called postmaturation dormancy.

Induced dormancy (secondary) develops after the dispersal or collection of nondormant seeds or seeds emerging with partial or total primary dormancy. Essentially, it reflects no sensitivity to germination inductors, internal or external. The main causes inducing dormancy in buried seeds are the level of humidity, the insufficiency or lack of light and oxygen, the presence of volatile or allelophatic inhibitors, and the high level of CO_2 (anaerobiosis or insufficient air). Germination can be inhibited by exposing the seeds to long periods of white light, especially to densities of high radiant flows or far-red light. Dormancy can also be prevented, delayed, or reduced by intermittent light of low intensity. Innate dormancy is absent in the recalcitrant seeds of the Tropics.

The induction of primary dormancy while the seed is on the parent tree, or the induction of secondary dormancy in the independent seed, can be complete or relative (partial). In a state of true or complete dormancy, the seed does not germinate under any environmental condition; if the dormancy is partial or relative, germination is restricted to a certain range of environmental conditions. Seed dormancy is subjected to constant changes; the increasing of dormancy is caused by cyclic changes following a seasonal pattern. The dormancy that persists when the seed returns to favourable environmental conditions is reinforced or induced.

The emergence from dormancy is frequently regulated by a promoter-inhibitor system, where the principal promoter is gibberellic acid (GA_3) and the main inhibitor is abscissic acid (ABA). Low levels of inhibitor and high levels of promoter induce germination. According to some studies, it is not possible at present to determine the precise function of ABA in the induction of dormancy.

The Seedling

The term seedling has not been precisely defined and its conceptualisation varies among authors. Seedling is defined as the youngest stage of the new sporophyte, from radicle protrusion through total liberation of the protective structures and abscission of the cotyledons, until the plant reaches a length of 50 cm. The numerous variations among seedlings of different species and the continuity among the programs of seed development-germination-seedling development in seeds lacking maturation drying make it difficult to establish the limits of this term. The definition provided establishes at least some minimal limits.

Seedling and Germination

Since Caesalpinius first defined seedling morphology in 1583, many have proposed different ways to classify seedlings. To alleviate confusion, to simplify the classification, and to deal with the numerous variations in tropical seedlings, two types of germination (epigeal and hypogeal) and two types of seedlings (phanerocotylar and cryptocotylar) are used. The types of germination refer to the process of germination while the seedling types emphasize cotyledon location.

Figure 1.

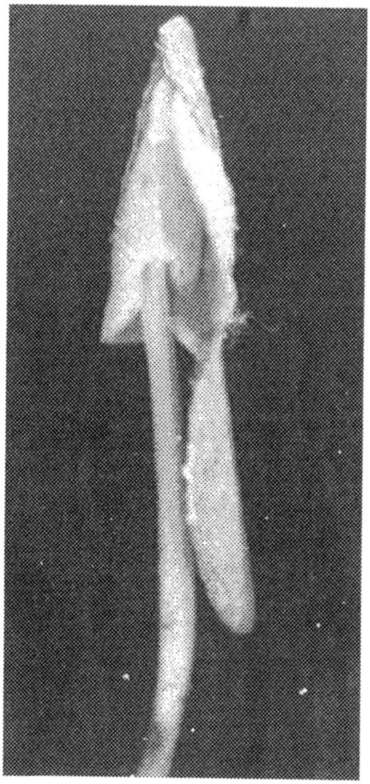

Figure 2

Figure 3.

Many seedlings have epigeal germination. The cotyledons (free or inside the seed, seed plus endocarp, or seed plus pericarp) and the cotyledonar node are affected by the distance to the soil level due to hypocotyl development. In hypogeal germination, the cotyledons and the cotyledonar node remain on the soil level, partially or totally immersed but seldom buried. The hypocotyl is very small or vestigial, sometimes not observable. In most cases, the cotyledons remain inside the seed. This type of germination is common in tropical trees and frequent in large, recalcitrant seeds.

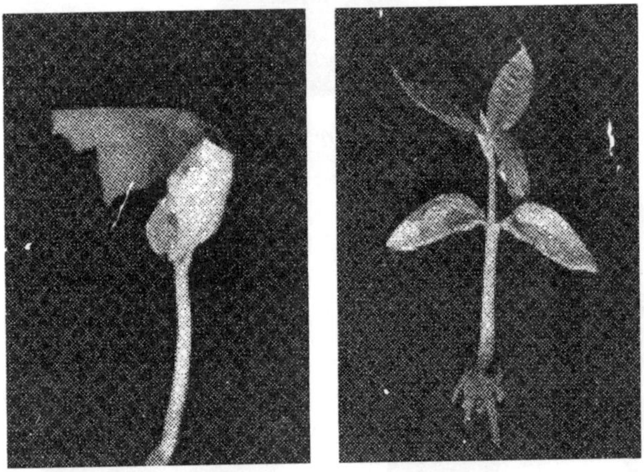

Figure 4 and 5

In the phanerocotylar seedling the cotyledons are outside the seed coat. The cotyledons are free. In the cryptocotylar seedling, the cotyledons remain enclosed in the seed coat (or seed coat plus endocarp, or seed coat plus pericarp). It does not matter whether they are large or small, storing or haustorial, free or fused, etc. This classification allows a combination of germination and seedling types.

Epigeal germination-phanerocotylar seedling

Examples include the seeds and seedlings of the following species: *Aca- cia, Adenanthera, Albizia, Cassia grandis, Dipteryx, Diphysa, Enterolobium, Erythrina, Gliricidia, Haematoxylum, Hymenaea courbaril, Hymenolobium, Parkia, Parkinsonia, Pterocarpus, Samanea saman, Sclerolobium, Tamarindus, Vatairea, Ormosia velutina, Stryphnodendron, Casuarina, Annona, Cymbopetalum, Mollinedia, Bernoullia, Cordia alliodora, Laetia, Psychotria, Simira maxonii, Palicourea, Guettarda, Genipa americana, Myrcia, Stemmadenia, Tetrapteryx, Vochysia, Qualea, Hyeronima, Capparis, Terminalia amazonia, T. oblonga, Guajacum sanctum, Cedrela, Melia, Zizyphus, Meliosma, Anacardium excelsum, Crescentia, Tabebuia, Jacaranda, Ulmus, Ilex, Casearia, Homalium, Rapanea, Dendropanax, Elaeocarpus, Vitex, Couratari, Couroupita, and Cariniana (figs. 1-5). Calatola costaricensis (Icacinaceae)* presents a problem of ubication.

The germination is epigeal, but initially the plumule remains enclosed within the seedcoat plus the endocarp. Later, the endocarp splits along two fissures and the two valves are dropped. The plumule continues its development inside the papyraceous seedcoat and continuously absorbs nutrients from the endosperm. The seedcoat remains intact until it explodes under the pressure of the expanding plumule. The plumule is several centimeters long and has a pair of large, green cotyledons and several young, developing leaves.

Epigeal germination-*cryptocotylar seedling*

Examples include *Virola, Otoba, Minquartia guianensis, Ximenia, Hura, Faramea, Omphalea, Sterculia apetala, and Durio* (figs. 6, 7). Hypogeal germination-phanerocotylar seedling-Examples are: *Allantoma, Lecythis, Barringtonia, Eschweilera, Grias, Bertholletia, Careya, and Corythophora* (the hypocoty remains inside the seedcoat; the cotyledons are two free cataphylls), *Inga, (destroyed sarcotesta), Garcinia (reduced, free cotyledons, sometimes seen as fleshy warts), and Caryocar (cotyledons free, scaly)* (figs. 8-10).

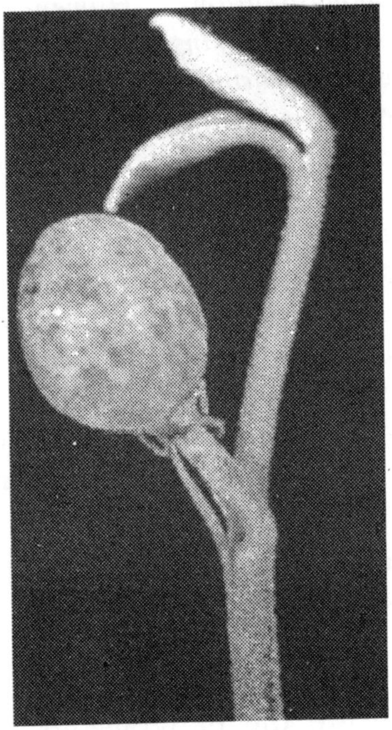

Figure 6.

Hypogeal germination-cryptocotylar seedling

Examples are *Calophyllum, Andira, Cynometra, Cojoba, Sophora, Spondias, Prioria, Mora oleifera, Myroxy on, Swartzia, Pentaclethra, Ocotea, Licaria, Nectandra, Persea, Swietenia, Carapa, Guarea, Trichilia, Brosimum, Poulsenia, Pseudolmedia, Mappia, Cupania, Gustavia, Eugenia, Syzygium, Lacmellea, Hernandia, Rourea, Hirtella, Chrysobalanus, Licania, Parinari, Prunus, Pachira aquatica,*

Sapindus saponaria, Melicoccus, Nephelium, Quararibea, Cavanillesia, Terminalia catappa, Pouteria, Quercus, Gynocaryum, Oreomunnea, and Alfaroa (figs 11).

Figure 7.

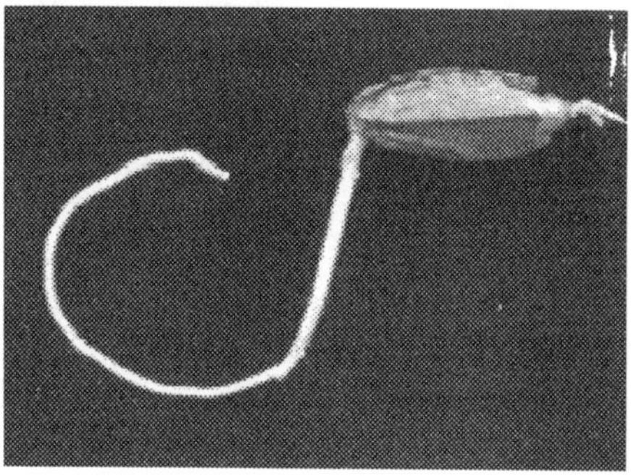

Figure 8.

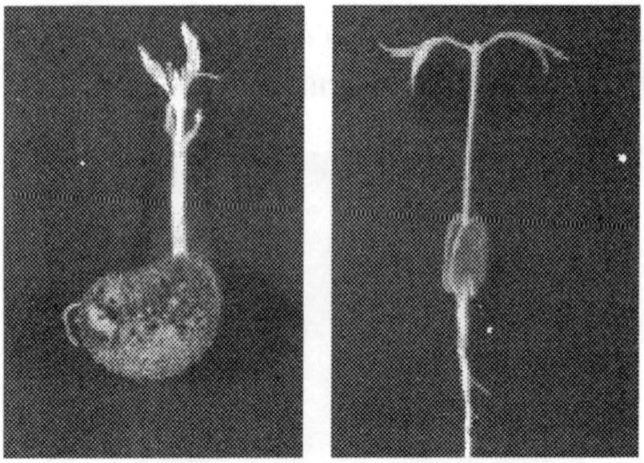

Figure 9 and 10

Seedling Structure

The seedling is formed by the root (radicle in the embryo); the hypocotyl; one, two, or several cotyledons (embryonic leaves); the epicotyl; and the leaves arising at the nodes of the epicotyl. The plumule, a term frequently used, is the embryonic shoot of the embryo or seedling, located above the cotyledonar node.

Figure 11 and 12

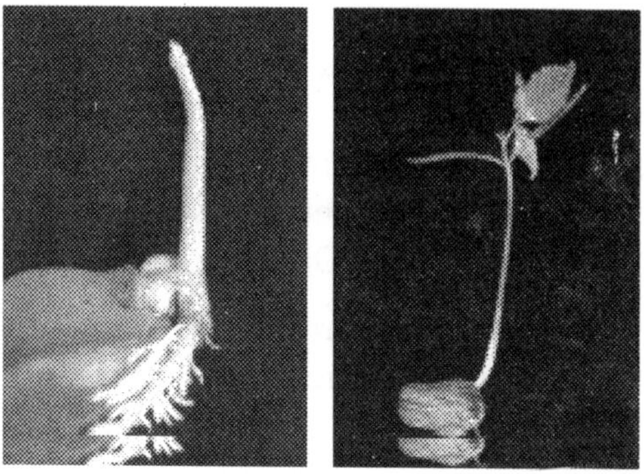

Figure 13 and 14

It includes the incipient epicotyl, the foliar primordia, and the apical meristem.

Root

The radicle of the embryo can be oriented facing the hilum (syntropous) or in the opposite direction (antitropous). In most seeds the radicle emerges through the micropyle; in some, the embryo pushes the seedcoat, fragments it, and protrudes through it at a specific point. This point or opening is called the operculum, embryostega, or plug. Opercula vary in ontogeny, structure, and opening mechanism. Opercula are formed by the micropylar and hilar regions, by the exostome and endostome, or by only the endostome. The operculum is more common in monocotyledons.

In numerous species, root emergence is more complicated because it requires going through the tissues of the seedcoat, the fruit, or the diaspore. This phenomenon is frequent in legumes with winged diaspores (Hymenolobium mesoamericanum, Sclerolobium, Myroxylon balsamum, Myrospermum frutescens, Pterocarpus hayesii, P. Officinalis Jacq.), indehiscent fruits with hard pericarp (Andira inermis, Prioria copaifera), drupes with a woody endocarp enclosing the seed(s) as a physical barrier (Dipteryx panamensis, Calatola costaricensis, Minquartia guianensis, Caryocar costaricense, Licania platypus), and diaspores in which the perianth pieces are hard and enveloping (Oreomunnea, Dipterocarpus).

In some seeds, root protrusion coincides with the liberation of characteristic odors and, sometimes, a change in seedcoat colour. The seedcoat of Balizia elegans (Fabaceae-Mimosoideae) is peculiar because it changes from the original greenish colour to purplish on the lateral surfaces as soon as the seed is imbibed, and also exudes a strong sulphur odor. The seedcoat of Schizolobium parahyba (Fabaceae-Caesalpinioideae) liberates tannins when root protrusion starts. In Stryphnodendron microstachyum and Abarema adenophora, the root leaches sulfuring substances, liberating a strong odor. The root of Abarema adenophora has nodules a few days after root protrusion.

The root is the organ that supports (anchors) the seedling and absorbs, conducts, and sometimes stores water and nutrients. Allorhizy is the process through which the seedling develops an embryonic root (radicle) as a pivotant or typical root. The first root is called the primary root. In the dicotyledons and the gymnosperms this root penetrates directly into the soil and gives rise to the secondary or lateral roots. The older roots are found near the root collar (zone of transition hypocotyl-root).

When a primary root is missing and the main root is adventitious, as happens in the pteridophytes, the seedling has primary homorhizy. In the monocotyledons and some dicotyledons, the primary root dies in early stages and the radical system is formed by adventitious roots forming a fibrous or fascicled root system in which no root dominates (secondary homorhizy). Typically, primary roots penetrate deeper in the soil than fibrous roots. Nevertheless, the superficiality of the fibrous roots and the toughness with which they adhere to the soil particles enable them to reduce soil erosion.

The root has several zones of growth-meristem or quiescent center, zone of cell division, region of elongation, and region of maturation. The radical hairs form in the maturation zone and their development can be inhibited by very wet or dry soils. The number of radical hairs varies among species and they can be absent in groups such as the Lauraceae (Ocotea, Nectandra, Persea) and Olacaceae (Minquartia guianensis). They have a short life, but in some woody species they may suberize or lignify, persisting for months or years. The center for the perception of gravity seems to be located in the root cap.

The extension of a radical system (depth and lateral extension) depends on factors such as humidity, temperature, and soil composition. Most absorbent roots are localised in the first meter of depth (juvenile forms and adults) and condensed in the upper 15 cm, which usually coincides with the richest organic matter. The lateral extension of the radical system usually coincides with that of the aerial system.

During growth, the balance between photosynthates (manufactured substances) and the water and minerals being absorbed by the roots is essential. During seedling establishment, the total of water and minerals absorbed generally exceeds that of the photosynthates produced. The balance is gradually reached with age, but transplanting seedlings always alters this process. The thinnest absorbing roots break when the plant is removed from the soil; pruning contributes to reestablishing the balance of shoot to root. Fungi, bacteria, and insects that attack the shoot or the root may also affect the balance between the shoot and the root.

Approximately 90 percent of the trees and, in general, the higher vascular plants, are associated with soil fungi (mycorrhizas). They have a symbiotic relationship. The two main categories established are the ectomicorrhizas and endomycorrhizas, which are based on the morphology of the mycorrhiza and the fungus involved. Seedlings lacking mycorrhizas grow slower and are frequently feeble. The mycorrhizic fungi with their extramatricial hyphae increase the survival and growth of the seedlings; the hyphae increase the area of root absorption for the intake of water and soil elements such as P, Cu, and Zn (especially P). The surface covered by the hyphae goes farther than that covered by the radical hairs. The mycorrhizic associations also protect against pathogens.

Seedlings with few radical hairs are highly dependent on mycorrhizas. Seedlings bearing mycorrhizas are also more tolerant to heavy toxic metals, pathogens attacking the roots, dryness, high soil temperatures, saline soils, adverse pH, and transplant. Frequently, species growing in marshes do not have mycorrhizas but when they grow in noninundated soil the mycorrhizic association becomes indispensable. The mycorrhizic rhizosphere is rich in bacteria, diatoms, and other fungi.

Mycorrhizic fungi depend on zoochorous dispersal. Animals may ingest sporocarps willingly or unwillingly, transporting the spores. The ingested spores maintain their viability through the digestive tracts of invertebrate and vertebrate vectors. Some propose that the spores' passage through the digestive tracts of mammals promotes germination. The sporocarps of diverse fungi are also part of the diet of numerous local fauna.

Other root associations common in the Tropics, and especially in legumes of the subfamilies Papilionoideae and Mimosoideae, are those established with nitrogen-fixing bacteria (bacterial nodules). The most common bacteria involved in root nodulation is Rhizobium. The bacterial symbiosis permits the fixation of atmospheric N and its conversion to ammonium, a soluble form of N that is easy to use and important for increasing forest productivity.

Numerous tree legumes have bacterial nodules, for example: *Abarema, Abrus, Acacia, Afzelia, Albizia, Anadenanthera, Andira, Baphia, Cedrelinga, Centrosema, Cojoba, Dalbergia, Detarium, Dialium, Dipteryx, Enterolobium Erythrina, Gliricidia, Indigofera, Inga, Lonchocarpus, Machaerium, Parkia, Pentaclethra, Piptadenia, Pithecellobium, Prosopis, Pterocarpus, Sclerolobium, Stryphnodendron, Swartzia, Tephrosia, and Platymiscium*. The presence of bacterial nodules has also been documented for Alnus, Coriaria, Elaeagnus, Hippophaë, Shepherdia, Comptonia, and Purshia. Some gymnosperms, such as Podocarpus, Libocedrus, Agathis, and Araucaria, also have bacterial nodules. Some semiparasitic species such as Ximenia (Olacaceae) have haustorial roots and some evidence exists of the parasitic behaviour of the roots of Minquartia guianensis (Olacaceae).

Hypocotyl

The hypocotyl is the part of the seedling axis that extends from the cotyledonar node downward to the root collar. The latter can be conspicuous or inconspicuous. Morphologically, it is the zone of transition between the root (with a radial distribution of alternating bundles of xylem and phloem) and the epicotyl, with vascular bundles formed by xylem and phloem. In the species with epigeal germination (phanerocotylar or cryptocotylar seedlings), the hypocotyl is well developed and its thickness, colour, and indumentum distinguish it from the root when the collar root is not conspicuous.

The seedlings with hypogeal germination (and phanerocotylar or cryptocotylar seedlings) usually have a small or vestigial hypocotyl (Calophyllum, Carapa, Cojoba, Pentaclethra, Quercus, Swietenia) that is barely observable (figs. 11, and 13). In several cases, the hypocotyl is a massive, storing organ, occupying the interior of the seed almost completely and remaining in the seed during germination. It remains hidden for weeks or months, thickens and enlarges slowly, and finally breaks the surrounding seedcoat or seedcoat and endocarp (Eschweilera, Lecythis, Caryocar). This last type of seed is recognised because the root and the epicotyl emerge at opposite ends of the seed.

Frequently, these seeds produce one or several adventitious roots at the epicotyl base or near the cotyledonar node. Garcinia and other Clusiaceae illustrate this kind of seedling development. However, Calophyllum (Clusiaceae) differs from this pattern. In this species, the cotyledons are fused in the distal three-quarters and seedling development follows the hypogeal-cryptocotylar pattern instead of the hypogeal-phanerocotylar pattern followed by other Clusiaceae.

Epicotyl

The epicotyl is the part of the seedling axis that is distal to the cotyledons (above the cotyledonar node). It develops from the activity of the apical meristem. It is well developed in many seeds of Fabaceae, Fagaceae, and other families. Usually, it is easily distinguished after germination and differs from the hypocotyl in thickness, texture, and colour.

Foliar Succession

Plants have four types of leaves: cotyledons, protophylls, metaphylls, and prophylls.

Cotyledons

The first leaf (or leaves) of the plant are the cotyledons. The most common condition is dicotyly, present in dicotyledons and some gymnosperms. Polycotyly (three or more cotyledons), syncotyly (gamocotyly or cotyledons partly or totally fused), schizocotyly (cotyledon division), pseudomonocotyly (false monocotyly due to cotyledonar fusion), monocotyly (one cotyledon), and anisocotyly (cotyledons of different sizes) represent the extremes of modification of the simple dicotyly. Polycotyly is quite common. Acer, Juglans, and Coffea frequently present three cotyledons. Numerous gymnosperms present the same phenomenon [e.g., Tsuga (2 to 7), Abies (2 to 10), Pseudotsuga, and Pinus (3 to 18)].

Qualea paraensis (Vochysiaceae) may present three or four cotyledons in some seedlings. Syncotyly is also common. For example, Calophyllum, Swietenia, Guarea, and Carapa have cotyledons fused distally. If the fusion is less than half the cotyledon length, they are called hemisyncotylous or amphisyncotylous. The presence of unequal cotyledons (anisocotyly) is frequent in such diverse seeds as those of Calophyllum, Peperomia, Brosimum, and other legumes.

Syncotyly is one of the primary causes impeding cotyledon emergence during the germination of seeds enclosed in hard seed coats, fruits with hard endocarps, or diaspores with other tissues attached. In these seeds the emergence of the epicotyl requires intercalary growth at the base of the cotyledons, leading to petiole formation. The elongation of the cotyledonary petioles displaces the cotyledonar node outside the seedcoat. Once outside, those petioles open backwards. Their concave adaxial surfaces provide space for the development of the plumule enclosed at the cotyledonary node base.

In species that are syncotylous, with hypogeal germination and a cryptocotylar seedling, the cotyledonary petioles are thick, sometimes short, woody or corky, and have a different color, but they are always adaxially channeled (figs. 11 and 13). If the germination is epigeal and the seedling is cryptocotylar, those petioles are longer, flexible, and with a softer texture (figs. 6-7). In some groups, the cotyledons have lobules. Sometimes the lobules are so deep that they suggest a possible schizocotyly.

Form, size, and thickness of cotyledons vary widely. Terms to describe cotyledon form include linear, reniform (Hymenolobium), cordate or auriculate (Bignoniaceae), elliptic (Guaiacum), lanceolate (Annona), obovate (Casuarina), oblong (Erythroxylum), scaly (cataphylls, Lecythis), trifoliate (Bursera simaruba), and oblong and planoconvex (Dipteryx). Many cotyledons have a petiole, but others are sessile or semisessile.

De Vogel proposed three types of cotyledons from a functional point of view: storing, haustorial, and photosynthetic during germination. Numerous cotyledosperm seeds have large, fleshy cotyledons, rich in stored reserves (e.g., Anacardium, Aspidosperma, Brosimum, Calophyllum, Carapa, Cojoba, Dipteryx, Enterolobium, Guarea, Hymenaea, Hymenolobium, Mangifera, Manilkara, Nectandra, Ocotea, Pachira,

Pentaclethra, Persea, Prioria, Quercus, and Sclerolobium). In many cases, the embryo stores nutrients in the cotyledons and the hypocotyl and the whole embryo is thick and fleshy.

The existence of haustorial cotyledons associated to hypogeal germination (cryptocotylar seedlings) was well documented by Heckel, who describes the germination of Ximenia americana (Olacaceae). This type of cotyledon is present in many families (Celastraceae, Olacaceae, Myristicaceae). In Myristica, Virola, and Otoba, the rudimentary embryo develops divergent cotyledonar blades that are large, very thin, and highly vascularised. The vascular bundles have numerous transfer cells for rapid transport. In Compsoneura, the cotyledons are haustorial but their structure is more complex. The Olacaceae (Ximenia americana, Minquartia guianensis) have minute embryos which develop haustorial cotyledons with a large blade, well vascularised and usually colored pink or violaceous (figs. 17-20).

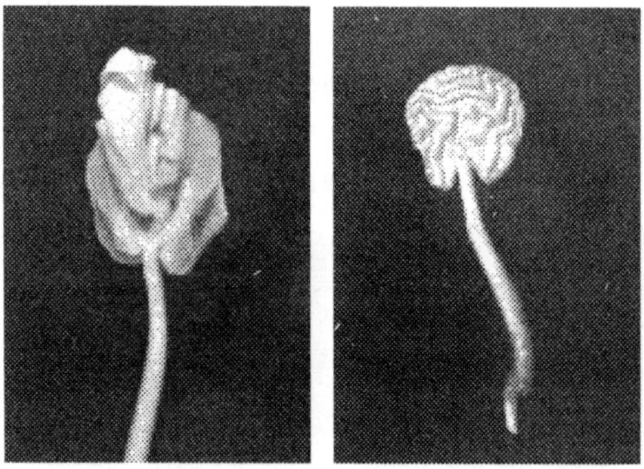

Figure 15 and 16

Green cotyledons are common in many seeds. Janzen found in his study of 74 species in the dry tropical forest of Costa Rica that 55 percent of the developing embryos from trees, shrubs, and herbs were chlorophyllous. The cotyledons, the radicle, and the hypocotyl were green. The seedcoats were white and transparent and the fruit pericarp was green. In seeds with maturation drying and dormancy, the embryos became nonchlorophyllous at the end of the maturation period.

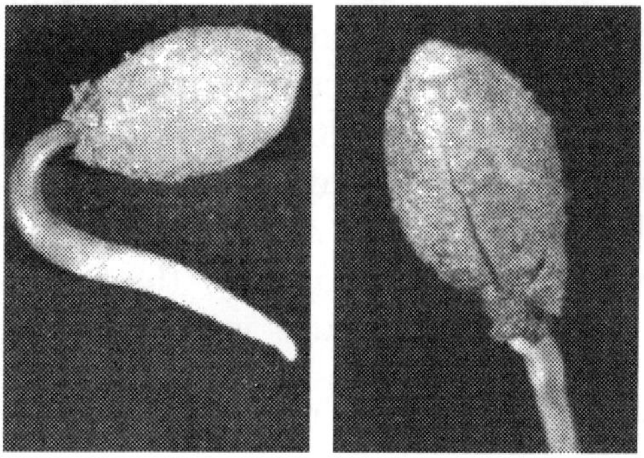

Figure 17 and 18

Janzen suggests that the chlorophyllous stage is important in embryo development and adds several variables to the complex interaction of fruit to seed(s), because fruit and seed form are selected to increase the exposure of the embryo to sunlight.

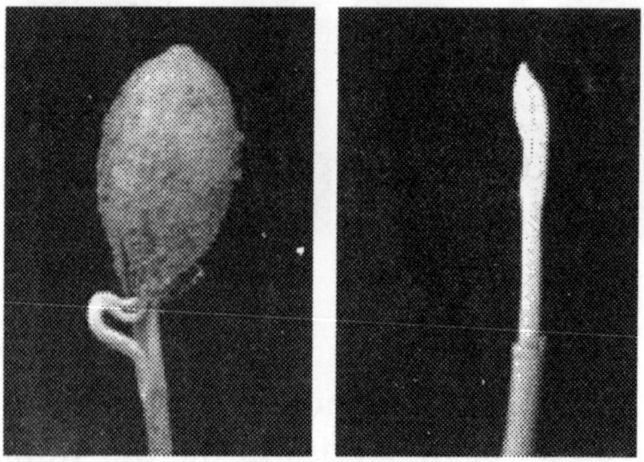

Figure 19 and 20

Astronium graveolens, Bursera simaruba, Terminalia oblonga, Ateleia herbert-smithii, Enterolobium cyclocarpum, Samanea saman, Gliricidia sepium, Haematoxylum basiletto H. Karst., Piscidia carthagenensis, Licania arborea, Calycophyllum candidissimum, and Simarouba glauca DC. are examples of trees that have seeds with chlorophyllous embryos. In some species (e.g., Inga, Dipteryx), the embryos with fleshy cotyledons are green during germination. In others, foliaceous cotyledons become chlorophyllous when germination begins. In many cases they increase in size (e.g., Vochysia , Qualea, Simira maxonii).

In Pentaclethra macroloba (hypogeal germination-cryptocotylar seedling), the cotyledons break the seedcoat and open like valves. The seedcoat remains attached to the abaxial surface of the cotyledons but the adaxial turns green and photosynthetic. These cotyledons have two functions: storage and production of photosynthates. When they assume the photosynthetic function, the cotyledons increase in size, change in texture, and develop stomata at the distal end of the cotyledon blade.

Simira maxonii (Rubiaceae) shows a different pattern. The embryo is spatulate and whitish, but when germination starts, the cotyledonar blade extends and turns green. As the cotyledonar blade expansion progresses, they begin to corrugate because the seedcoat limits their extension. Finally, the pressure they exert on the seedcoat causes it to break and the cotyledons become free (figs. 15-16).

In most species, the cotyledonar buds may develop if the epicotyl is damaged by herbivory or other factors. This phenomenon is frequent in Meliaceae (Carapa, Guarea), Fabaceae (Cojoba, Pentaclethra), and Fagaceae (Quercus). However, some species have more than a single axillary bud (e.g., Swietenia, Carapa, and Pentaclethra macroloba). The cotyledon axils and the leaves of P. Macroloba have descendent series of accessory buds, which develop in a predetermined order to generate a shoot that assumes apical dominance. These buds have ecological significance because they guarantee a high rate of seedling survival even in the limiting conditions imposed by predation and other factors in the low strata of the forest.

A peculiar case requiring further research is the capacity of the cotyledons of Gustavia superba (Lecythidaceae) to regenerate. The cotyledons removed from young seedlings were able to regenerate root and shoot. This capacity represents an advantage for the species because rodents predate the fruits and seeds and the seedlings, like most Lecythdaceae seedlings, are exposed to a high rate of herbivory on the forest floor.

Protophylls

These are the juvenile leaves of the plant. In seedlings, they are called cataphylls, or eophylls.

Cataphylls

This term, derived from the Greek cataphyllum, means the lower leaves. The cataphylls are the inferior leaves, between the cotyledons and the eophylls. They are reduced forms, simpler than the metaphylls and, in general, are squamiform, membranaceous, hyaline or coriaceous, and brown. Usually, they lack chlorophyll. The scales of the invernating buds or perules, as well as those of the subterranean stems, are also called cataphylls. Cataphylls are frequently found in the epicotyl of many plants (e.g., Carapa, Swietenia, Pentaclethra, Lecythis, Eschweilera, Calophyllum,

Pseudolmedia, and Mappia). Their size, shape, and structural complexity usually increase along the axis in an acropetal direction. In many species they drop off early in the plant's development.

Eophylls

The term means precocious or juvenile leaves. Tomlinson defined eophylls as the first leaves with a green and expanded blade. In many seedlings, one or several cataphylls precede the eophylls. Families such as the Anacardiaceae, Bignoniaceae, Fabaceae (Caesalpinioideae, Mimosoideae, Papilionoideae), Meliaceae, and Sapindaceae have seedlings with a transitional series of leaves between the eophylls and the metaphylls. The variation (heterophylly) in the series can be constant or can vary among seedlings of the same species. Abrupt change and lack of seriation is typical in species such as Guajacum sanctum (Zygophyllaceae).

The seedlings of this species and Hymenolobium mesoamericanum (Fabaceae-Papilionoideae) have paripinnate eophylls, similar to the metaphylls. Certain other species, such as Bursera simaruba, have trifoliate cotyledons, simple eophylls, and complex metaphylls. Still others, such as Tilia, have lobed cotyledons followed by eophylls and metaphylls with simple blades. Some seedlings have eophylls that are more dentate or lacerate than the metaphylls (Cordia, Casearia, Hasseltia, Poulsenia, Quercus, and Rapanea).

Metaphylls

These are the typical leaves of the adult plant, usually morphologically different from protophylls. The typical leaf has a blade or limb, a foot or petiole, and a foliar base. The petiole can be missing (sessile leaf) or can be short and reduced (semisessile leaf). The petiole participates in the transport of substances, supports the foliar blade, and situates it in a position favourable to receiving sunlight. In several species the blade does not develop and the petiole may bear thorns, creeping organs, or tendrils. In some species, laminar petioles (phyllodes) substitute for the foliar blade. Acacia heterophylla and A. Mangium Willd. Show a series of transitions from leaf to phyllode. Leaves can be simple or compound. In simple leaves, the blade does not divide into minor units.

In compound leaves, the blade divides into folioles or pinnae. The foliar blade varies in size, form, texture, and other characteristics. Shapes range from linear to circular, with a great richness of variation in blade, margin, base, and apex. Families such as the Casuarinaceae (Allocasuarina, Casuarina) have whorled metaphylls, gamophyllous (congenital concrescence), reduced, and scaly. The gymnosperms have several types. The best known is the acicule (needle) typical of Pinus. In other gymnosperms the leaves are larger and have different shapes. The Podocarpaceae have lanceolate, linear, or oval leaves. The cycads have pinnate leaves; Bowenia have bipinnate, and those of Ginkgo biloba are fan-shaped.

In compound leaves, the folioles or pinnae may grow from a central point, as in Tabebuia or Ceiba, forming palmatecompound leaves; if they arise in pairs along a rachis, as in Spondias, Simarouba, Enterolobium, and Samanea saman, the leaf is pinnate-compound. The leaf with a terminal pinna is imparipinnate; if a terminal pinna is missing, the leaf is paripinnate. When the compound leaf has more complex subdivisions, the secondary axes arising at the rachis are called rachillae and the small foliar blades are called foliolules or pinnules.

The folioles or pinnules can be sessile, semisessile, or have a petiolule. In the petiole base a basal thickening called pulvinus can be found; the petiolule can have a pulvinulus. In some species, the petiole has a double pulvinus (e.g., Sloanea, Elaeocarpaceae). The petiolule can have double pulvinuli as in Protium (Burseraceae). The pulvini and pulvinuli are common in Fabaceae and are related to nyctinasty.

The texture or consistency of the leaf blade can be cartilaginous (hard, tough, flexible), crustaceous (hard, thin, brittle), coriaceous (hard, thick, leathery), chartaceous (papery, opaque, thin), membranous (thin, semitranslucent, membrane-like), or hyaline (thin, translucid). Some cotyledons and leaves lack trichomes and are called glabrous; others have indument or vestiture. Those with trichomes can be pubescent (hairs or trichomes short and soft), pilose (soft, shaggy trichomes), puberulent (minutely pubescent, like dust), velutinous (dense, straight, long and soft trichomes), floccose (dense, appressed trichomes in patches or tufts), hirsute (long, rather stiff trichomes), hispid (very long, stiff trichomes), lanate (long, intertwined trichomes), papillose (with papillae), sericeous (long, silky trichomes, commonly appressed), strigose (sharp, coarse, bent trichomes with wide bases), tomentose (dense, interwoven trichomes). The type of trichome can be used for taxonomic purposes. The indument can be ferruginous, whitish, grayish, golden, reddish, or colourless.

The cotyledons and metaphylls can be stipulate or exstipulate. The stipules are lateral appendages emerging from both sides of the foliar base. Their development often precedes the foliar blade development. They can be connected to the foliar base (adnate) or be free. Most are small and scaly, but some are foliaceous, or modified as thorns (Robinia, Acacia), and some form glands. When the leaves are opposite or whorled, the contiguous stipules fuse and form an interpetiolar stipule as in Rubiaceae (Coffea, Randia, Hamelia, Cinchona, and Psychotria), or an ocrea (ocreate base) by congenital concrescence and longitudinal elongation as in Polygonaceae. The deciduous stipules leave stipular scars on both sides of the leaf base. When the stipules are narrow and have an acute apex they are lacinate.

Prophylls

The prophylls are the first leaves of the lateral buds giving rise to branches. The dicotyledons have a pair of prophylls transverse to the branch axis; the monocotyledons have one parallel to the principal axis. The pair of prophylls in the dicotyledons are usually opposite and can have reduced forms, sometimes squamiform. In adult trees, some species with floral branches have concrescent prophylls that may form special structures such as the cupule in Fagaceae.

Leaf Venation

The leaf venation is formed by a system of vascular bundles or veins distributed throughout the whole leaf blade in close relation with the mesophyll. The venation of the leaf blade has taxonomic value. The leaves of most dicotyledons have a midvein and a network of veins progressively smaller, arranged in a reticular pattern. Close inspection of the minor veins reveals that the areole or islets-small areas of mesophyll limited by intersecting veins-contain small, single, or ramified veins ending freely in the mesophyll. The following classification is modified from Hickey.

Pinnate

A midvein is the origin of the major lateral veins.

Craspedodromous

All secondary veins end at the leaf margin. It can be simple (all of the secondary veins and their branches ending at the margin), semi-craspedodromous (secondary branches within the margin, one branch terminating at the margin, the others joining the superadjacent secondary), or mixed (a combination of the two former subtypes).

Camptodromous

The secondary veins do not end at the margin. These are of four types.

- *Brochidodromous*: The secondary veins are joined together in a series of prominent arches, forming a pseudo-marginal vein.
- *Eucamptodromous*: The secondary veins arch upwards, gradually diminish distally inside the margin, connecting to the superadjacent secondary veins through a series of cross veins without forming conspicuous marginal loops.
- *Reticulodromous*: The secondary veins lose their identity towards the leaf margin due to repeated branching, forming a reticulum.
- *Kladodromous*: The secondary veins branch freely toward the margin.

Hyphodromous

Only a primary vein (midvein) is present. It can be rudimentary or concealed.

Striatodromous

This type was called parallelodromous by Hickey; however, as in the venation described by Troll for the monocotyledons, the major veins arising at the foliar base are not equidistant along their course and are progressively anastomose toward the leaf apex. This type of venation was called closed striate by Troll.

Campylodromous

Several primary veins or their branches arise in or near a single point and extend upwards, forming recurved arches before converging at the leaf apex.

Acrodromous

Two or more well-developed primary or secondary veins extend in convergent arches toward the leaf apex. Their origin at the leaf base can be basal or suprabasal; the development of the lateral veins can be perfect or imperfect.

Actinodromous

Three or more primary veins diverge radially from a single point.

Palinactinodromous

The primary veins have one or more subsidiary points of radiation above the lowest point. They can be basal or suprabasal. The development can be perfect (marginal or reticulate), imperfect (marginal, reticulate) or flabellate.

The most common type in the leaves of the monocotyledons is the closed striate. In some families the lateral veins diverge and form a striate-arcuate pattern, in others a pinnatestriate pattern. The acicular leaves of gymnosperms have one or two vascular bundles. In others with pinnate leaves, the pinnae can have a single vein as in Cycas, a pinnate venation as in Stangeria, or a ramified dichotomous venation as in Ceratozamia mexicana. Some species, such as Ginkgo biloba, Circaeaster, and Kingdonia, have dichotomous venation.

The cotyledons may have any of the former types; frequently, their venation is simpler and can differ from the type found in the metaphylls. The eophylls may have intermediate or transitional types.

Chronological and Morphological Variations in Seedling Development

The timing of visible and invisible (inside seed before root emerges) events of seedling development vary among species. The following five examples show the numerous chronological and morphological variations that exist in seedling development, and although general patterns are established, each seed and each seedling is unique.

Dipteryx panamensis has a large embryo and the plumule is well developed. Visible germination (root emergence) begins at 10 days. The two endocarpic valves open along the dorsal and ventral sutures and the seedcoat splits along the raphe-antiraphe line. The greenish hypocotyl begins its development at 13 to 15 days and elevates the plumule which is enclosed inside the partially open valves. The eophylls are compound and pinnate with stipellate folioles. Germination is epigeal, and the seedling is phanerocotylar.

Calophyllum brasiliense has a large, well-developed embryo, with cotyledons that are gamocotylous in the distal three-quarters and concave adaxially in the basal quarter. The plumule is enclosed within this small cavity.

The embryo axis is poorly developed. Visible germination begins between 16 and 18 days, and plumule emergence requires the formation and extension of cotyledonar petioles. The hypocotyl is vestigial. The epicotyl produces several deltoid cataphylls before developing the eophylls. Germination is hypogeal, and the seedling is cryptocotylar.

Virola koschnyi and Minquartia guianensis have a minute embryo and a massive endosperm, rich in lipids; in V. koschnyi the endosperm is ruminate. In both species, the embryo has a long, intense period of development inside the seed before the root emerges. Both embryos develop haustorial cotyledons and cotyledonar petioles that propel the plumule. Both species have cryptocotylar seedlings. Although the process is similar in both species, V. koschnyi seedlings develop faster.

In both, germination is epigeal and the seedling is cryptocotylar. Lecythis ampla seedlings develop very slowly. The embryo is large and the entire seed cavity is filled by a massive hypocotyl. The radicle and the plumule are reduced to a small mass of meristematic cells at opposite ends of the hypocotyl. The radicle forms 35 to 40 days after sowing, and the root emerges at 45 days. Sometimes an adventitious root emerges before the primary root. Plumule development begins at 7 months and is very slow. The cotyledons are green, squamiform structures. Several cataphylls are produced before the eophylls. Germination is hypogeal, and the seedling is phanerocotylar.

Seedling Ecology

Seedling ecology is only partially known and many factors affect seedling development and survival. These factors include spectral composition of light, presence of shade, type of soil, extent of herbivory, and association of bacteria and fungi. The spectral composition of light affects seed germination and seedling development. In the forest, the canopy trees modify the spectral composition because the leaves absorb the photosynthetically active radiation (400 to 700 nm) but allow the near-infrared radiation (700 to 3000 nm) to penetrate to the lower strata. The red far-red radiation ratio (R/FR) decreases with clearings at 1.22, forest gaps at 0.90, sunflecks in the understory at 0.87, and diffuse light in the understory at 0.40. The canopy also provides differences in temperature and moisture in the different microenvironments of the forest.

The seedlings of many tropical trees do not tolerate shade (heliophytes) and develop only in forest clearings and gaps; others are indifferent to shade and may develop under full shade (tolerant, sciophytes). Some are partially tolerant to shade (partial sciophytes); these can survive under shade but need light to develop fully. The heliophytes of the natural forest can be ephemeral or durable. The ephemeral are pioneers with a short life; they have light, soft wood and rapid growth, and occupy forest clearings.

Approximately 10 years later, durable heliophytes begin to dominate. These species live longer and produce better wood. The partial sciophytes are established next, followed by the sciophytes. The sciophytes grow slower and live longer than partial sciophytes. The incidence of shadeintolerant species is high in the different strata of the forest and increases toward the canopy. Shade-intolerant species comprised 71 percent of the species in the canopy, 48 percent in the subcanopy, and 39 percent in the understory in La Selva, Sarapiquí, Costa Rica.

Although seedling behaviour in the forest is very important to consider for natural forest management and forest regeneration, the vastly different ecology of the plantation evokes very different and often unexpected responses, which often limit success. Twenty-three native species planted in experimental blocks near a community of San Miguel de Sixaola, Talamanca, Costa Rica, showed that the seedlings of some species did not survive under the shade of the forest; some experienced high mortality rates under full sunlight; and others were indifferent to light level and quality, flourishing in all treatments. Under full sunlight, most seedlings show higher longitudinal growth and higher increments in diameter.

Species such as Dalbergia retusa, Dipteryx panamensis, Rollinia pittieri, Virola koschnyi, Genipa americana, Terminalia amazonia, and T. Oblonga grew better under full sunlight; under moderate shade, the juveniles were smaller and thinner but still survived well. Cordia alliodora was the only species that grew best under partial shade; however, it did not survive under full shade. Virola sebifera was the only species that grew well under partial or total shade.

Many of the species that grow best under full sunlight in plantations behave as partial sciophytes in the natural forest. Under natural conditions, they cannot compete with climbers, weeds, or other pioneer trees or shrubs in open spaces; in plantations, competition is reduced by periodic ground clearings three to four times per year). In the forest, the species survive well under partial shade but their growth is slower (in height and diameter). Heliophytes, on the other hand, compete with climbers and weeds by natural pruning (autopruning), growing a dense crown and exfoliating bark early. These differences demonstrate why seedling behaviour must be studied in the natural forest, the greenhouse, the nursery, and the different types of plantations.

Soil type, especially in natural habitats, also influences seedlings. Some grow well in acid (oxylophytes) or saline (halophytes) soils, others in sandy (psammophytes) or rocky (lithophytes, chasmophytes) soils. The differences in soil can affect seed germination; seedling vigor; the size, strength, and woodiness of the stem; the depth of the radical system; the pubescence; the susceptibility to dryness, cold, or pathogens; and, eventually, the flowering and fortification of the mature tree.

Seedlings can be deeply influenced by minor differences in soil composition produced by forest clearing or by the local effect of the canopy trees on the soil. Some species have very specific soil requirements; for example, Vochysia ferruginea Mart., V. guatemalensis and V. allenii Standl. & L.O. Williams may form monospecific stands or patches with Dipteryx panamensis and Pentaclethra macroloba in clayey, acid soils (pH 5.0 to 6.0), with high concentrations of iron and bauxite.

Herbivory plays an essential role in maintaining diversity in the tropical forest . Herbivores reduce seed establishment under the maternal tree through intense predation. The greater the density of the seedlings under the conespecific adult trees, the greater the predation by herbivores and pathogens. This density-dependent mortality can be common to many species in the forest, but the strategies of growth vary among species and affect the structure of the population.

In species such as Xylopia brasiliensis (Annonaceae), variations in phenological patterns and fruit availability define the forage patterns, which in turn affect the patterns of seed dispersal. The real influence of herbivory on species diversity in the forest can be determined only by studying the process dynamics; the restriction of the census to spacing patterns is insufficient.

In the plantations of several neotropical forest tree species, herbivory intensifies and seedlings come under greater fungal attack. Colletotrichum sp. Produces the derrite of the apex in Dipteryx panamensis; Phomopsis, the foliar spots of Virola koschnyi leaves; and Nectria, the seedling death of Stryphnodendron microstachyum. Rodents and other mammals in forests and plantations can also predate seedlings (e.g., deer eat the shoots of seedlings of Dipteryx panamensis, provoking stem bifurcation).

The influence of bacteria (nodules) and fungi (mycorrhiza) associations on seedling survival and development is only partially known. John proposes that mycorrhizic symbiosis promotes seedling development and increases transplant survival. Because symbiotic fungi are probably lacking in the areas being restored, inoculation of seedlings may be the best alternative. Future research on these associations would increase information about what bacteria and fungi to use in tropical plantations.

5

Physiological Features of Seedling

Seedling quality will be defined as fitness for purpose, which for a seedling is its ability to survive and then grow rapidly when planted in the field. Both morphological and physiological features are undoubtedly required to meet this objective. The ease with which most morphological parameters can be measured makes them the most popular method for measuring seedling quality.

For example, all nursery grading standards are based on morphology. In spite of general acceptance and widespread use, researchers have noted that morphology alone does not predict all the variability seen in field survival and growth. It can, however, be of great comparative value when the physiological status of the seedlings is equal (i.e., in a single lot of seedlings from a given nursery) or similar (i.e., different nurseries but similar lifting dates).

Measurement of Height

One of the easiest morphological traits to observe is height. No instruments are necessary to visually determine that one tree is taller than another. If you wish a quantitative measurement, height is usually measured from the root collar to the base of the terminal bud. If no terminal bud is present as a result of injury or active growth, measurement is done either to the highest point (often the tip of a needle) or to the approximate growing point. The growing point method gives a more accurate measure and does not result in height fluctuations when sequential measurements are taken.

Standards

In the past, seedlings have been culled both for being too tall and too short. Exceedingly tall seedlings are difficult to plant, out of balance (poor shoot-root ratio) and susceptible to wind rock. Tall seedlings may, however, be genetically superior. Rather than culling these seedlings, top-mowing or clipping has been instituted in the nursery to reduce height and make these seedlings plantable. Top-mowing generally produces a more uniform crop with fewer culls due to extreme seedling size. Minimum standards for height vary greatly by species, seed-zone, and age class. Standards are generally revised as more information on field performance becomes available from the field.

Quality Prediction

Height measures both the photosynthetic capacity of the seedlings and their transpirational area by being highly correlated to needle number. This would suggest a good correlation of height with growth but an unpredictable relationship with survival, especially on droughty sites. Indeed, height has not been consistently correlated with survival.

In West Germany tall Norway spruce seedlings had poorer survival than smaller ones but had greater subsequent growth. Similarly, the tallest seedlings again showed the poorest survival for Douglas-fir and

Ponderosa pine. Other studies show little difference or increasing survival with height for Douglas-fir. Increased height growth with increasing initial height is a more universal phenomenon. In red pine, initial seedling sizes of 7, 11, and 15 cm resulted in 299, 311, and 366 cm tall trees, respectively, after twelve years in the field.

Loblolly pine shows even more striking growth differences based on initial height which result in larger total differences as the seedlings grow. The effect of initial height on subsequent height is seen to vary with stock-type for white spruce. The predictive power of height as a measure of field survival and growth is best summarised by Mullin and Svaton for white spruce.

As initial seedling height increases the height of the white spruce trees after ten years in the field increases, as does the total growth in the field. Survival is not linearly related to initial height but shows a definite maximum. The maximum is probably dependent on seedling root system size and site related. As the site becomes more droughty, the optimum height for survival probably decreases.

Measurement of Stem Diameter

Diameter is often included in the morphological grading done at the nursery. Diameter should be measured slightly above the root collar with either an accurate caliper for research purposes or a caliper gauge for nursery grading. For accurate measurements, care should be taken to compress the bark as little as possible and to exert a constant pressure.

Standards

Most nurseries employ only a lower limit for acceptable diameter. As with height standards, they vary greatly with species and stock-type.

Quality Prediction

Seedling diameter is related to field performance in the same manner as height. It is generally accepted as a better measure of both growth and survival. This may simply be a reflection of the flatter survival curve for diameter when compared to height and the tendency of the curve to flatten out at proportionally larger diameters. As with height, diameter is not always correlated to field survival but is related to subsequent growth.

For example, loblolly pine seedlings graded to Wakeley's standards in two separate studies had the best survival at the middle grade. After 13 years in the field, however, the seedlings in the Blair and Cech study had produced 6.5, 6.0, and 5.0 cu. ft. of wood per tree for grades 1, 2, and 3, respectively. The relationship between initial stem diameter and six-year height for two stock-types of spruce and white pine was highly significant but differed by stocktype. Likewise, radiata pine field growth and survival after three years were well-correlated with initial stem diameter. A quality seedling should possess the largest diameter that confers an acceptable level of survival potential on that seedling for a given site.

Measurement of Weights

Measurement of weight can be done on either a fresh weight or dry weight basis. Fresh weights, although easier to measure, vary greatly with the water content of tissue. For this reason, greater consistency is achieved with dry weight measurements. The most common tissues measured are either whole tree, or root, shoot, and leaf, separately. Tissue should be washed thoroughly before drying. If separate shoot and root dry weights are desired, the seedlings are severed at the root collar or some repeatable position and each half is placed in a separate paper bag. The temperature for drying should be high enough to denature the enzymes responsible for decomposition (> 60°C) while not causing thermal decomposition and nitrogen volatilisation (< 70°C). Therefore, 65°C is recommended especially if any tissue nutrient analysis is anticipated. After coming to a constant weight, usually 24 hours, tissue is removed from the bag and weighed.

Standards

Dry weight is not used as a grading criterion because such measurement is both time-consuming and destructive. Weight measurements are generally used to evaluate the suitability of a lot of seedlings rather than individual seedlings. Additionally, weights are often used to calculate shoot-root ratios and a quality index.

Quality Prediction

A strong correlation exists between seedling dry weight and stem diameter for many species such as loblolly pine and Douglas-fir. Dry weight correlates to field survival and growth with the same level of consistency as diameter. In addition to total dry weight predictive value, Hulten reported a close correlation between foliage dry weight of oneyear-old Scots pine seedlings and their height in the second and third year. Since height is correlated to needle number and needle number to needle weight, needle weight would be expected to correlate to growth and survival in the same way as height. In general, a quality seedling should be as heavy as possible to produce the best growth while still having the balance of shoot and root necessary to survive on a given site.

Measurement of Common Shoot Parameters

In addition to the commonly measured parameters, a number of other factors can be examined. Some of these predict field growth (bud height), while others may indicate a lack of vigor (color, foliage density), or improper physiological condition (color, succulent shoots). A final group of standards examined in this section is used to remove stock that may be damaged in the nursery (swept stem).

Bud Height or Length

The height or length of the bud is easily measured in many species that have large- or mediumsized winter resting buds. To measure the height, it is easiest to remove the bud just below the first formed bud scales with a sharp razor blade. Calipers can then be used to measure the height. Care must be taken so that the calipers do not crush the bud. The size of the bud and the number of needle primordia it possesses are dependent on when the buds were formed and how vigorous the seedlings were as they became dormant.

Similarly, bud height is correlated to the number of needle primordia in Douglas-fir and to subsequent field growth in both determinant growth species, such as Ponderosa pine and red pine, and the multiflushing species, loblolly pine and slash pine. Bud height may be an indicator that will probably be more useful for potential field growth comparisons between lots of seedlings rather than within a lot. It is an indirect method of measuring time of dormancy induction and seedling vigor. The strong correlations with field height make it an important morphological indicator of growth potential.

Measuring Seedling Color

Seedling color is often used as a subjeccive measure of seedling quality. In a gross sense, it probably has value. Yellow, brown, or light green seedlings are less vigorous than dark green ones. Within the range of dark green color, however, the evidence for correlations with field growth and survival is less convincing. During a given season, more foliar nitrogen results in more chlorophyll and, therefore, in a darker green color. The relative survival potential of the greener tree depends to some extent on the relative nitrogen content.

In black spruce, nitrogen contents above 2.2% in the leaves, although producing a greener tree, result in a substantial decrease in frost hardiness. Van den Driessche found that nitrogen concentrations in the seedlings at outplanting correlated to survival for coastal Douglas-fir and sitka spruce and to height growth for coastal and interior Douglasfir and sitka spruce but that no relationships were found with lodgepole pine growth or survival.

In southern pine container-grown seedlings, chlorophyll content was inconsistently correlated with field survival but proved a good indicator of subsequent field growth. By contrast, color of white spruce seedlings at time of planting was not related to either survival or growth after ten years in the field. Although the evidence is varied, dark green color is probably essential to a quality seedling.

Secondary Needles

Secondary needles are those needles that appear in pine in bundles surrounded by a sheath (fascicular) during the first or second season of growth. It has generally been accepted that the presence of these needles is a good indicator of seedling development in southern pine. Secondary needle development occurs when the stem is woody and more resistant to stress, indicating an ability to withstand planting shock.

Evidence suggests that this relationship may not hold in all pine species. Thompson compared two types of 1-0 container-grown Scots pine seedlings; one group with normal 1-0 field morphology (very few secondary needles and a terminal rosette of needles surrounding the bud) and a second group resembling 2-0 Scots pine (many secondary needles and a terminal bud with surrounding lateral buds). The findings indicate that seedlings with the "normal" 1-0 morphology start growth slightly later in the spring but consistently produce larger seedlings in the field even with similar initial heights and calipers.

Thompson concludes that seedlings with the 1-0 type morphology in the greenhouse are more desirable because of their greater growth potential. Presence of secondary needles on container- grown pine seedlings must be evaluated on each species before its value as a morphological measure of quality can be determined. 6.5.4 Succulent Shoots vs. Dormant Buds Lemmas or late season growth often produces shoots that remain succulent at lifting. In Douglas-fir, survival of seedlings with succulent shoots was lower than for those without, but the difference was not significant. Although the succulent shoots increase seedling height, late season growth delays budset and delays the normal sequence of fall dormancy induction. This, in turn, delays spring bud break. Seedlings demonstrating delayed bud burst generally have decreased growth and survival in the field.

In Northeast species, actively growing stock is easily broken and dies or wilts when outplanted under water stress. Seedlings without active growth at the time of lifting may be one universally acceptable grading criterion for improved seedling growth and survival in the field.

Needle Size and Density

Needle length and the distance between needles decrease when seedlings are grown under water stress conditions (bottle brush growth). Nursery transplants and seedlings with root diseases often exhibit this type of growth. For transplants, a short section of growth of this type is probably of no consequence, but seedlings with longer sections (most of last year's growth) should be removed at grading. Needle density can be decreased by changing bed density. Thompson found that in red pine higher bed densities resulted in increased elongation of stem units with no additional stem units produced, resulting in greater height but identical photosynthetic area. Needle density also may reflect the light condition under which the seedlings were grown. Loblolly pine seedlings grown under 2/3 full light had fewer needles per unit of stem than those grown under full light. An optimum foliage density presumably exists for various species. Extremely high foliage density or short needle lengths should probably be avoided for quality seedlings.

Multiple shoots

Multiple shoots can either be genetic in nature or result from cultural practices (top-mowing), disease, or insects. Many Northwest nurseries cull for multiple tops that split within a given distance from the ground line. This is intended to eliminate genetically determined multiple tops. When multiple tops result from damage or insects in the nursery, the seedlings—usually recover rapidly in the field and produce a single terminal. Some reduction in growth may occur during the first year while side shoots compete for dominance.

Stem sweep

Stem sweep often occurs when seedlings are transplanted

incorrectly and grow to right themselves. How much sweep is too much and what effect, if any, it has on field survival and growth are unknown. Sweep, however, is often used as a grading criterion for transplants.

Disease deformities

Disease can produce some pronounced stem deformities. Phomopsis canker can girdle Douglas-fir stems, producing a large bulge above the girdle. Fusiform rust in southern pine also produces a characteristic stem deformity. All diseased seedlings should be removed in grading.

Root Size Measurement

Although the root is at least as important as the shoot in determining survival and growth of the plant, no quantitative measure exists that is wholly satisfactory. Root mass, root volume, root length and root area index are common measurements. Root mass and volume do not give an accurate representation of root fibrosity because seedlings with many fine roots may weigh or displace the same amount as one with a large tap root. Total root length is a better measure of fibrosity or absorptive surface but it is too tedious and time-consuming to be practical. Root area index is determined photometrically and estimates root density and distribution within the soil profile.

Standards

Grading is done on root size using very subjective standards of sufficient root for the height and visual density.

Quality Prediction

Root weight is often correlated to seedling diameter and

thereby to survival and growth. Although root weight correlated to growth and survival of container loblolly pine seedlings, height, diameter and stem weight were all better predictors. In studies where fibrosity was intentionally reduced, or seedlings visually graded before planting, field results differ. In lodgepole pine and Douglas-fir, root fibrosity correlated well with field survival. Loblolly pine seedlings had large increases in mortality when 50 to 75% of the feeder roots were removed. Lavender and Wareing, on the other hand, abraded roots and removed fibrous roots of Douglas-fir but found little difference in subsequent growth and survival. Visual grading of roots in two additional studies, also, failed to correlate with survival and growth.

These types of comparisons from the literature might lead one to believe that root quality is unimportant. This is probably misleading. The problem more likely is that we do not have a good measure by which to compare root systems and we have not quantified how much root is necessary to assure survival and growth upon outplanting. A quality seedling is one that possesses a large root system, comprised of a high percentage of fibrous roots. The implication is that a fibrous root system has a large surface area for absorption of water and nutrients.

Root Parameters

Root Deformity

Roots serve to both anchor the seedling and to explore the soil volume for water and nutrients. If the natural habit of the root is disturbed in an early stage by confinement in a container or manipulation in the nursery, the root growth form can be permanently altered, sometimes with detrimental effects.

Container shape is important in determining the number of roots and the degree to which those roots spiral in the container. Spiralling roots may strangle the tree, reducing growth and ultimately causing death. Seedlings grown in containers for a longer time than desirable may become pot-bound; this can slow their growth both in the container and the field.

Seedlings may, also, only have active root tips at egress holes at the bottom of the container. When outplanted, few if any new roots originate in the top 10 cm of the plug. These seedlings may not be wind-firm and may be damaged or uprooted by strong winds. A new method of treating containers with cupric carbonate may reduce this problem.

In bare-root nurseries, undercutting by running a sharp blade under the seedlings in the bed is a common practice. Schmidt-Vogt reports that undercutting can affect older fir seedlings by inhibiting their subsequent tap root development. He reports 10-year-old trees toppling in the wind and believes more attention should be paid to this area in the future.

Root sweep in transplants, although avoidable, is nevertheless common. How much sweep is permissible and its effect on growth and survival are debatable. In general, Northwest seedlings are culled if the main axis of the root system is greater than 45° from that of the stem. In outplanting trials, root sweep was associated with reduced growth of Norway spruce. Loblolly pine root systems, deformed to simulate planting problems, accumulated carbohydrates at the bend, which resulted in more roots developing at that point. This would probably be desirable on mesic sites but detrimental on droughty sites.

After several years in the field, a quality seedling root system should resemble, as much as possible, that of a field-grown tree. How the root system should look at planting to achieve this goal is not fully known.

Stiff Lateral Roots

Stiff lateral roots that originate near the root collar and extend over 10 cm in any direction are difficult to plant in the field. To alleviate this problem, seedlings in the nursery beds are root-pruned between the seedling rows. Some seedlings with stiff laterals still remain; these are generally culled.

Mycorrhizae

Mycorrhizae is the association of the seedling root with a symbiotic fungus. In conifers, it is often visible as a fuzzy mantle on the roots. Characteristic bifurcate roots (branching) are also formed. The author knows of no grading system that uses mycorrhizae as a criterion.

Some mycorrhizal associations have been shown to confer superior growth and survival potential to seedlings under certain conditions. For example, Pisolithus tinctorius on loblolly pine roots has resulted in up to a 25% increase in both growth and survival when the seedlings were planted on poor sites. Less dramatic results are seen on better sites. A quality seedling should have a beneficial mycorrhizal association to insure the greatest likelihood of superior survival and growth. This requirement is probably more important when seedlings are destined for harsh sites.

Root Damage

In addition to seedlings culled for insufficient roots, seedlings with sufficient but damaged roots are also culled. Root damage can be caused by insects such as weevils or cranberry girdlers that eat the bark at or below the root collar or by diseases such as Phytophora and Black root rot which are seen as black, soft roots. The most common form of damage is created during the lifting process, when roots are ripped off leaving gashes or severely abraded roots. Seedlings with damaged roots are seen as poor risks in the field and should be removed.

Morphological Indices

A morphological index is a combination of two or more morphological measurements. They are generally designed to serve one of two purposes. The first is to describe an abstract attribute of a seedling, such as balance and sturdiness. The second is to determine the relative importance of the morphological measurements by combining them into an index that more closely corresponds to field performance than any individual parameter.

Shoot-root Ratio

Measurement

The shoot-root ratio (S/R) was devised as a measure of balance between the transpirational area (shoot) and the water absorbing area (root) of a seedling. Racey et al. gave a thorough account of the methods of measuring S/R and how they compare for various populations. In general, either dry weight or volume displacement of the shoot and root are used to obtain the ratio. As the S/R ratio increases, the volumetric measure increasingly overestimates the dry weight ratio. The error is of little concern, however, for the S/R range generally encountered in seedlings.

Quality prediction

Much controversy exists over the value of S/R in

determining seedling quality. When seedlings were first graded into height classes, Hermann (using Douglas-fir) and Lopushinsky and Beebe (with both Douglas-fir and ponderosa pine) found that seedlings with lower S/R within a height class had better survival. It has been demonstrated that S/R changes with seedling size. Therefore to determine if treatments affect S/R or to find an acceptable S/R for a seedling, the effect of size must be removed. To do this, S/R for the Douglas-fir seedlings in the two studies mentioned above were plotted against total seedling height. This exercise demonstrates that as seedling height increases, the "acceptable" S/R also increases. Wrenching studies in which the S/R was altered have shown mixed results; in some cases, large changes in survival were reported while in other cases the influence on survival was small. Bed density was found to significantly change seedling S/R but to have little effect on survival.

In Southern pine, the value of S/R in predicting quality may be more difficult to assess. Seedlings grown in warmer climates can continue to grow roots as long as the soil temperature allows. The S/R ratio, in turn, changes rapidly and dramatically during the lifting season. one study with seedlings grown in Virginia reports a change in S/R for loblolly pine from 5.5 in October to 3.8 in January and finally to 2.8 in March.

This reflects an actual change in S/R due to root dry weight increase at the expense of shoot dry weight with little change in total dry weight. In another study with loblolly pine in southern Georgia, the seedlings accumulated dry weight during the winter and the S/R ratio changed from approximately 1.0 in October to 0.5 in March. These study results do not correspond to allometric relationships proposed for loblolly pine.

In the past several years, the usefulness of S/R has been questioned by many nursery managers. S/R should be re-evaluated against height or dry weight, especially for species that do not reallocate their dry weight or grow substantially during the winter. Within limits, it appears that S/R can be a useful index for predicting field survival but probably has little value in predicting field growth. A quality seedling should have as low a shoot/root ratio as possible to insure the best survival.

Sturdiness Quotient

The sturdiness quotient is the height (h) in centimeters divided by the stem diameter (d) in millimeters (h/d). It reflects the stocky or spindly nature of the seedlings. Although a good indicator of the ability to withstand physical damage in all stock-types, it is of particular importance in container-grown seedlings where the sturdiness quotient can get very high on undesirable spindly stock. Roller found that black spruce seedlings with sturdiness quotients greater than six were seriously damaged when exposed to wind, drought, and frost. In general, sturdiness quotient should closely parallel diameter in predicting survival and growth in the field.

Dickson Quality Index

$$\text{Quality Index} = \frac{\text{Total seedling Dry wt(g)}}{\dfrac{\text{Height (cm)}}{\text{Diameter (mm)}} + \dfrac{\text{Shhot wt (g)}}{\text{Root wt (g)}}}$$

The quality index was devised by evaluating how well a number of possible combinations of morphological parameters predicted field performance of white spruce and white pine seedlings and selecting the best combination. In a subsequent test, this index was able to predict quality based on the nutrient environment (soil fertility) in which the seedlings were grown. The index was successfully used by Roller to differentiate between plantable and non-plantable containerised seedlings. In Douglas-fir, the index may reflect the outplanting success of various stock-types.

Iyer and Wilde Quality Index

The index proposed by Iyer and Wilde combines diameter-height ratio, root-shoot ratio, specific gravity of the stem, catalytic capacity of the feeder roots, and the coefficient of variability of the height into a single number. The determination of this number is somewhat complex, but the authors claim success in correctly determining the quality of stock from four nurseries. The method needs further testing to assess its reliability.

Relative Predictive Value of Morphological Parameters

Diameter is probably the best and easiest-tomeasure overall predictor of subsequent growth and survival. It must be remembered that past a certain size, survival begins to decline as diameter increases. This may reflect a lack of balance in the larger tree.

Shoot-root ratio corrected for seedling size is another good indicator of survival potential. It may be of particular importance on droughty sites. The index may have value in determining when seedlings become too large and out of balance. When seedling heights or diameters are equal for two lots of trees but their bud heights are different, those with the larger buds have the greatest field growth potential. In this way, bud height can be used to refine predictions of field growth.

The remaining morphological features may have value as grading criteria to remove damaged or poorly-grown trees, but their usefulness as predictors of field performance is either unproven or unlikely. Some grading procedures, such as height or diameter segregation, put a directional genetic selection pressure on the seedling population, while other grading standards exert little genetic influence.

6

Morphological Evaluation of Seedling

Seedling quality will be defined as fitness for purpose, which for a seedling is its ability to survive and then grow rapidly when planted in the field. Both morphological and physiological features are undoubtedly required to meet this objective. The ease with which most morphological parameters can be measured makes them the most popular method for measuring seedling quality. For example, all nursery grading standards are based on morphology. In spite of general acceptance and widespread use, researchers have noted that morphology alone does not predict all the variability seen in field survival and growth.

It can, however, be of great comparative value when the physiological status of the seedlings is equal (i.e., in a single lot of seedlings from a given nursery) or similar (i.e., different nurseries but similar lifting dates). There are several parameters routinely used to grade seedlings, to evaluate the effect of grading on subsequent field performance, and to show that morphology is an important indicator of seedling survival and growth in the field.

Height

Measurement

One of the easiest morphological traits to observe is height. No instruments are necessary to visually determine that one tree is taller than another. If you wish a quantitative measurement, height is usually measured from the root collar to the base of the terminal bud. If no terminal bud is present as a result of injury or active growth, measurement is done either to the highest point (often the tip of a needle) or to the approximate growing point. The growing point method gives a more accurate measure and does not result in height fluctuations when sequential measurements are taken.

Standards

In the past, seedlings have been culled both for being too tall and too short. Exceedingly tall seedlings are difficult to plant, out of balance (poor shoot-root ratio) and susceptible to wind rock. Tall seedlings may, however, be genetically superior. Rather than culling these seedlings, top-mowing or clipping has been instituted in the nursery to reduce height and make these seedlings plantable. Top-mowing generally produces a more uniform crop with fewer culls due to extreme seedling size. Minimum standards for height vary greatly by species, seed-zone, and age class. Standards are generally revised as more information on field performance becomes available from the field.

Quality Prediction

Height measures both the photosynthetic capacity of the seedlings and their transpirational area by being highly correlated to needle number. This would suggest a good correlation of height with growth but an unpredictable relationship with survival, especially on droughty sites. Indeed, height has not been consistently correlated with survival.

In West Germany tall Norway spruce seedlings had poorer survival than smaller ones but had greater subsequent growth. Similarly, the tallest seedlings again showed the poorest survival for Douglas-fir and Ponderosa pine. Other studies show little difference or increasing survival with height for Douglas-fir.

Increased height growth with increasing initial height is a more universal henomenon. In red pine, initial seedling sizes of 7, 11, and 15 cm resulted in 299, 311, and 366 cm tall trees, respectively, after twelve years in the field.

Loblolly pine shows even more striking growth differences based on initial height which result in larger total differences as the seedlings grow. The effect of initial height on subsequent height is seen to vary with stock-type for white spruce. Chavasse reports that in radiata pine the initial height difference are maintained but not enlarged after one yea in the field.

The predictive power of height as a measure of field survival and growth is best summarised by Mullin and Svaton for white spruce. As initial seedling height increases the height of the white spruce trees after ten years in the field increases, as does the total growth in the field. Survival is not linearly related to initial height but shows a definite maximum. The maximum is probably dependent on seedling root system size and site related. As the site becomes more droughty, the optimum height for survival probably decreases.

A quality seedling with respect to height is one that is as tall as possible while still possessing an acceptable level of survival potential for the designated site.

Stem Diameter

Measurement

Diameter is often included in the morphological grading done at the nursery. Diameter should be measured slightly above the root collar with either an accurate caliper for research purposes or a caliper gauge for nursery grading. For accurate measurements, care should be taken to compress the bark as little as possible and to exert a constant pressure.

Standards

Most nurseries employ only a lower limit for acceptable diameter. As with height standards, they vary greatly with species and stock-type.

Quality Prediction

Seedling diameter is related to field performance in the same manner as height. It is generally accepted as a better measure of both growth and survival. As with height, diameter is not always correlated to field survival but is related to subsequent growth.

Weights

Measurement

Measurement of weight can be done on either a fresh weight or dry weight basis. Fresh weights, although easier to measure, vary greatly with the water content of tissue. For this reason, greater consistency is achieved with dry weight measurements.

The most common tissues measured are either whole tree, or root, shoot, and leaf, separately. Tissue should be washed thoroughly before drying. If separate shoot and root dry weights are desired, the seedlings are severed at the root collar or some repeatable position and each half is placed in a separate paper bag. The temperature for drying should be high enough to denature the enzymes responsible for decomposition (> 60°C) while not causing thermal decomposition and nitrogen volatilisation (< 70°C). Therefore, 65°C is recommended especially if any tissue nutrient analysis is anticipated. After coming to a constant weight, usually 24 hours, tissue is removed from the bag and weighed.

Standards

Dry weight is not used as a grading criterion because such measurement is both time-consuming and destructive. Weight measurements are generally used to evaluate the suitability of a lot of seedlings rather than individual seedlings. Additionally, weights are often used to calculate shoot-root ratios and a quality index.

Quality Prediction

A strong correlation exists between seedling dry weight and stem diameter for many species such as loblolly pine and Douglas-fir. Dry weight correlates to field survival and growth with the same level of consistency as diameter. In addition to total dry weight predictive value, Hulten reported a close correlation between foliage dry weight of oneyear-old Scots pine seedlings and their height in the second and third year. Since height is correlated to needle number and needle number to needle weight, needle weight would be expected to correlate to growth and survival in the same way as height.

In general, a quality seedling should be as heavy as possible to produce the best growth while still having the balance of shoot and root necessary to survive on a given site.

Common Shoot Parameters

In addition to the commonly measured parameters, a number of other factors can be examined. Some of these predict field growth (bud height), while others may indicate a lack of vigor (color, foliage density), or improper physiological condition (color, succulent shoots). A final group of standards examined in this section is used to remove stock that may be damaged in the nursery (swept stem).

Bud height or length

The height or length of the bud is easily measured in many species that have large-or mediumsized winter resting buds. To measure the height, it is easiest to remove the bud just below the first formed bud scales with a sharp razor blade. Calipers can then be used to measure the height. Care must be taken so that the calipers do not crush the bud.

The size of the bud and the number of needle primordia it possesses are dependent on when the buds were formed and how vigorous the seedlings were as they became dormant. Similarly, bud height is correlated to the number of needle primordia in Douglas-fir and to subsequent field growth in both determinant growth species, such as Ponderosa pine and red pine, and the multiflushing species, loblolly pine and slash pine.

Bud height may be an indicator that will probably be more useful for potential field growth comparisons between lots of seedlings rather than within a lot. It is an indirect method of measuring time of dormancy induction and seedling vigor. The strong correlations with field height make it an important morphological indicator of growth potential.

Colour

Seedling colour is often used as a subjeccive measure of seedling quality. In a gross sense, it probably has value. Yellow, brown, or light green seedlings are less vigorous than dark green ones. Within the range of dark green color, however, the evidence for correlations with field growth and survival is less convincing.

During a given season, more foliar nitrogen results in more chlorophyll and, therefore, in a darker green color. The relative survival potential of the greener tree depends to some extent on the relative nitrogen content. In black spruce, nitrogen contents above 2.2% in the leaves, although producing a greener tree, result in a substantial decrease in frost hardiness.

Van den Driessche found that nitrogen concentrations in the seedlings at outplanting correlated to survival for coastal Douglas-fir and sitka spruce and to height growth for coastal and interior Douglasfir and sitka spruce but that no relationships were found with lodgepole pine growth or survival. In southern pine container-grown seedlings, chlorophyll content was inconsistently correlated with field survival but proved a good indicator of subsequent field growth. By contrast, color of white spruce seedlings at time of planting was not related to either survival or growth after ten years in the field. Although the evidence is varied, dark green color is probably essential to a quality seedling.

Secondary Needles

Secondary needles are those needles that appear in pine in bundles surrounded by a sheath (fascicular) during the first or second season of growth. It has generally been accepted that the presence of these needles is a good indicator of seedling development in southern pine.

Secondary needle development occurs when the stem is woody and more resistant to stress, indicating an ability to withstand planting shock.

Evidence suggests that this relationship may not hold in all pine species. There are two types of 1-0 container-grown Scots pine seedlings; one group with normal 1-0 field morphology (very few secondary needles and a terminal rosette of needles surrounding the bud) and a second group resembling 2-0 Scots pine (many secondary needles and a terminal bud with surrounding lateral buds).

Succulent Shoots and Dormant Buds

In Douglas-fir, survival of seedlings with succulent shoots was lower than for those without, but the difference was not significant. Although the succulent shoots increase seedling height, late season growth delays budset and delays the normal sequence of fall dormancy induction. This, in turn, delays spring bud break. Seedlings demonstrating delayed bud burst generally have decreased growth and survival in the field.

In Northeast species, actively growing stock is easily broken and dies or wilts when outplanted under water stress. Seedlings without active growth at the time of lifting may be one universally acceptable grading criterion for improved seedling growth and survival in the field.

Needle Size and Density

Needle length and the distance between needles decrease when seedlings are grown under water stress conditions (bottle brush growth). Nursery transplants and seedlings with root diseases often exhibit this type of growth. For transplants, a short section of growth of this type is probably of no consequence, but seedlings with longer sections (most of last year's growth) should be removed at grading.

Needle density can be decreased by changing bed density. Thompson found that in red pine higher bed densities resulted in increased elongation of stem units with no additional stem units produced, resulting in greater height but identical photosynthetic area. Needle density also may reflect the light condition under which the seedlings were grown. An optimum foliage density presumably exists for various species. Extremely high foliage density or short needle lengths should probably be avoided for quality seedlings.

Multiple Shoots

Multiple shoots can either be genetic in nature or result from cultural practices (top-mowing),, disease, or insects. Many Northwest nurseries cull for multiple tops that split within a given distance from the ground line. This is intended to eliminate genetically determined multiple tops. When multiple tops result from damage or insects in the nursery, the seedlings—usually recover rapidly in the field and produce a single terminal. Some reduction in growth may occur during the first year while side shoots compete for dominance.

Stem Sweep

Stem sweep often occurs when seedlings are transplanted incorrectly and grow to right themselves. How much sweep is too much and what effect, if any, it has on field survival and growth are unknown. Sweep, however, is often used as a grading criterion for transplants.

Disease Deformities

Disease can produce some pronounced stem deformities.

Phomopsis canker can girdle Douglas-fir stems, producing a large bulge above the girdle. Fusiform rust in southern pine also produces a characteristic stem deformity. All diseased seedlings should be removed in grading.

Root Size

Measurement

Although the root is at least as important as the shoot in determining survival and growth of the plant, no quantitative measure exists that is wholly satisfactory. Root mass, root volume, root length and root area index are common measurements. Root mass and volume do not give an accurate representation of root fibrosity because seedlings with many fine roots may weigh or displace the same amount as one with a large tap root. Total root length is a better measure of fibrosity or absorptive surface but it is too tedious and time-consuming to be practical. Root area index is determined photometrically and estimates root density and distribution within the soil profile.

Standards

Grading is done on root size using very subjective standards of sufficient root for the height and visual density.

Quality Prediction

Root weight is often correlated to seedling diameter and thereby to survival and growth. Although root weight correlated to growth and survival of container loblolly pine seedlings, height, diameter and stem weight were all better predictors. In studies where fibrosity was intentionally reduced, or seedlings visually graded before planting, field results differ. In lodgepole pine and Douglas-fir, root fibrosity correlated well with field survival. Loblolly pine seedlings had large increases in mortality when 50 to 75% of the feeder roots were removed.

A quality seedling is one that possesses a large root system, comprised of a high percentage of fibrous roots. The implication is that a fibrous root system has a large surface area for absorption of water and nutrients.

Common Root Parameters

Root Deformity

Roots serve to both anchor the seedling and to explore the soil volume for water and nutrients. If the natural habit of the root is disturbed in an early stage by confinement in a container or manipulation in the nursery, the root growth form can be permanently altered, sometimes with detrimental effects.

Container shape is important in determining the number of roots and the degree to which those roots spiral in the container. Spiralling roots may strangle the tree, reducing growth and ultimately causing death. Seedlings grown in containers for a longer time than desirable may become pot-bound; this can slow their growth both in the container and the field.

Seedlings may, also, only have active root tips at egress holes at the bottom of the container. When outplanted, few if any new roots originate in the top 10 cm of the plug. These seedlings may not be wind-firm and may be damaged or uprooted by strong winds. A new method of treating containers with cupric carbonate may reduce this problem.

In bare-root nurseries, undercutting by running a sharp blade under the seedlings in the bed is a common practice. Schmidt-Vogt reports that undercutting can affect older fir seedlings by inhibiting their subsequent tap root development. Root sweep in transplants, although avoidable, is nevertheless common. How much sweep is permissible and its effect on growth and survival are debatable. In general, Northwest seedlings are culled if the main axis of the root system is greater than 45° from that of the stem.

In outplanting trials, root sweep was associated with reduced growth of Norway spruce. Loblolly pine root systems, deformed to simulate planting problems, accumulated carbohydrates at the bend, which resulted in more roots developing at that point. This would probably be desirable on mesic sites but detrimental on droughty sites. After several years in the field, a quality seedling root system should resemble, as much as possible, that of a field-grown tree. How the root system should look at planting to achieve this goal is not fully known.

Stiff Lateral Roots

Stiff lateral roots that originate near the root collar and extend over 10 cm in any direction are difficult to plant in the field. To alleviate this problem, seedlings in the nursery beds are root-pruned between the seedling rows. Some seedlings with stiff laterals still remain; these are generally culled.

Mycorrhizae

Mycorrhizae is the association of the seedling root with a symbiotic fungus. In conifers, it is often visible as a fuzzy mantle on the roots. Characteristic bifurcate roots (branching) are also formed. The author knows of no grading system that uses mycorrhizae as a criterion. Some mycorrhizal associations have been shown to confer superior growth and survival potential to seedlings under certain conditions. For example, Pisolithus tinctorius on loblolly pine roots has resulted in up to a 25% increase in both growth and survival when the seedlings were planted on poor sites.

A quality seedling should have a beneficial mycorrhizal association to insure the greatest likelihood of superior survival and growth. This requirement is probably more important when seedlings are destined for harsh sites.

Root Damage

In addition to seedlings culled for insufficient roots, seedlings with sufficient but damaged roots are also culled. Root damage can be caused by insects such as weevils or cranberry girdlers that eat the bark at or below the root collar or by diseases such as Phytophora and Black root rot which are seen as black, soft roots. The most common form of damage is created during the lifting process, when roots are ripped off leaving gashes or severely abraded roots. Seedlings with damaged roots are seen as poor risks in the field and should be removed.

Morphological Indices

A morphological index is a combination of two or more morphological measurements. They are generally designed to serve one of two purposes. The first is to describe an abstract attribute of a seedling, such as balance and sturdiness. The second is to determine the relative importance of the morphological measurements by combining them into an index that more closely corresponds to field performance than any individual parameter.

Shoot-root Ratio

Measurement

The shoot-root ratio (S/R) was devised as a measure of balance between the transpirational area (shoot) and the water absorbing area (root) of a seedling. Racey et al. gave a thorough account of the methods of measuring S/R and how they compare for various populations. In general, either dry weight or volume displacement of the shoot and root are used to obtain the ratio. As the S/R ratio increases, the volumetric measure increasingly over-estimates the dry weight ratio. The error is of little concern, however, for the S/R range generally encountered in seedlings.

Quality Prediction

Much controversy exists over the value of S/R in determining seedling quality. When seedlings were first graded into height classes, Hermann (using Douglas-fir) and Lopushinsky and Beebe (with both Douglas-fir and ponderosa pine) found that seedlings with lower S/R within a height class had better survival. It has been demonstrated that S/R changes with seedling size. Therefore to determine if treatments affect S/R or to find an acceptable S/R for a seedling, the effect of size must be removed. To do this, S/R for the Douglas-fir seedlings in the two studies mentioned above were plotted against total seedling height. This exercise demonstrates that as seedling height increases, the "acceptable" S/R also increases.

Bed density was found to significantly change seedling S/R but to have little effect on survival. In an attempt to explain the contradictory results found when S/R is used to predict survival, the results from the comparable wrenching and bed density studies mentioned above were plotted versus seedling height. In Southern pine, the value of S/R in predicting quality may be more difficult to assess. Seedlings grown in warmer climates can continue to grow roots as long as the soil temperature allows.

In the past several years, the usefulness of S/R has been questioned by many nursery managers. S/R should be re-evaluated against height or dry weight, especially for species that do not reallocate their dry weight or grow substantially during the winter. Within limits, it appears that S/R can be a useful index for predicting field survival but probably has little value in predicting field growth.

A quality seedling should have as low a shoot/root ratio as possible to insure the best survival.

Sturdiness Quotient

The sturdiness quotient is the height (h) in centimeters divided by the stem diameter (d) in millimeters (h/d). It reflects the stocky or spindly nature of the seedlings. Although a good indicator of the ability to withstand physical damage in all stock-types, it is of particular importance in container-grown seedlings where the sturdiness quotient can get very high on undesirable spindly stock. Roller found that black spruce seedlings with sturdiness quotients greater than six were seriously damaged when exposed to wind, drought, and frost. In general, sturdiness quotient should closely parallel diameter in predicting survival and growth in the field.

Dickson Quality Index

The quality index was devised by evaluating how well a number of possible combinations of morphological parameters predicted field performance of white spruce and white pine seedlings and selecting the best combination. In a subsequent test, this index was able to predict quality based on the nutrient environment (soil fertility) in which the seedlings were grown. The index was successfully used by Roller to differentiate between plantable and non-plantable containerised seedlings. In Douglas-fir, the index may reflect the outplanting success of various stock-types.

Iyer and Wilde Quality Index

The index proposed by Iyer and Wilde combines diameter-height ratio, root-shoot ratio, specific gravity of the stem, catalytic capacity of the feeder roots, and the coefficient of variability of the height into a single number. The determination of this number is somewhat complex, but the authors claim success in correctly determining the quality of stock from four nurseries. The method needs further testing to assess its reliability.

Relative Predictive Value of Morphological Parameters

Diameter is probably the best and easiest-tomeasure overall predictor of subsequent growth and survival. It must be remembered that past a certain size, survival begins to decline as diameter increases. This may reflect a lack of balance in the larger tree. Shoot-root ratio corrected for seedling size is another good indicator of survival potential. It may be of particular importance on droughty sites. The index may have value in determining when seedlings become too large and out of balance.

When seedling heights or diameters are equal for two lots of trees but their bud heights are different, those with the larger buds have the greatest field growth potential. In this way, bud height can be used to refine predictions of field growth. The remaining morphological features may have value as grading criteria to remove damaged or poorly-grown trees, but their usefulness as predictors of field performance is either unproven or unlikely.

Some grading procedures, such as height or diameter segregation, put a directional genetic selection pressure on the seedling population, while other grading standards exert little genetic influence.

7

Principles and Procedures of Root Growth Potentials

One of the most widely used, and perhaps most important, tests of seedling quality is the root growth potential (RGP) test developed in the 1950s. The test can be conducted using simple, inexpensive materials and generally gives results which are repeatable and easily interpreted.

RGP is the ability of a seedling to initiate and/or elongate roots when held in an environment which is conducive to root growth. Surprisingly, seedlings do not always grow new roots in such an environment. A seedling which does is said to have high RGP and can be expected to be of high vigour and performance (survival and growth) potential. High RGP is an important seedling quality attribute presumably because it enables the seedling to become established rapidly after planting. The rationale for this is that when a seedling is planted it has a finite root system.

Although it is capable of exploiting moisture and nutrients in its immediate vicinity, these reserves are soon depleted. So for establishment to occur new soil reserves must be tapped, hence new roots must be grown.

Seedlings which are unable to grow roots rapidly are doomed to water stress and, ultimately, death.

Development Factors for RGP

A seedling develops RGP while it is growing in the nursery or greenhouse but RGP is not expressed until the seedling is planted. The internal (endogenous) and external (exogenous) factors which influence these processes.

Planting stock characteristics such as species, seed lot, family and stock type play a key role in the intensity of RGP that the stock develops. Lodgepole pine (*Pinus contorta* Dougl.), for example, can developer greater RGP than can Englemann spruce (*Picea engelmanii* Parry) grown under the same nursery conditions. Ponderosa pine seedlings from seed zone 721 (near Klamath Falls) have consistently poor RGP from year to year and tend to show high mortality as well.

Coastal Douglas-fir (*Pseudotsuga menziesii* (Mirb) Franco) seedlings from seed zone 070-15 (near Coos Bay) have generally high RGP, while those from 030-05 (near Gray's Harbor) have poor RGP. Sutton reported large differences in RGP among several lots of jack pine (*Pinus banksiana* Lamb.) and black spruce (*Picea mariana* (Mill.) .S.P.) seedlings. Even heritable between family differences have been reported in *Pinus radiata* D. Don. Far more examples could be cited.

Other key factors influencing RGP are the physiological condition (esp. status of winter dormancy) of the seedling at the time it is lifted and possibly the size of the carbohydrate pool available for root growth. The former is a factor over which the grower can exercise considerable control. Growers can also control the length of time during which stock is held in cold storage. This interacts with lifting date to impact RGP at planting time.

In general, seedlings lifted in mid-winter tend to have higher RGP than fall- or spring-lifted seedlings and also have higher RGP following storage. Storage temperature is also important since temperatures above 0°C permit the proliferation of storage mold which tends to weaken seedling vigour and RGP. The condition of the seedling shoot and foliage is also important for RGP. Since leaves of many tree species export an essential rooting co-factor, removal of, or damage to, the foliage can impede root growth. So a seedling which has suffered defoliation by insects, frost or other agents in the nursery would be expected to have low RGP.

Expression of RGP

Recalling that RGP represents only a potential to grow roots, it follows that this potential may or may not be fully expressed when the seedling is outplanted. If it is planted into an environment optimal for root growth it will, by definition, fully express its RGP. If not, it will not. Since planting site conditions in winter and early spring are rarely optimum for root growth, it is probably rare for RGP to be fully expressed after outplanting. Perhaps the most important limiting factor to RGP expression is soil temperature. The optimum for seedling root growth for many species is near 20 degrees.

Soil moisture is another key factor. Tree seedlings are very sensitive to soil moisture stress. This is exhibited as a reduction in root growth when stress is imposed after planting. In one experiment Douglasfir seedlings were potted in forest soil and held under four different watering regimes for 30 days. Controls were maintained at near field capacity. Soil in the low, moderate and high stress regimes was permitted to dry down to -1, - 2 and -6 bars by the end of the period. These data show that soil moisture stresses approaching -2 bars depressed RGP to about 1/3 of the control value. Similar data are available for other species. Other factors which probably depress RGP expression are soil compaction, poor planting and other agents, which are reviewed in Ritchie and Dunlap.

Measuring RGP

The basic procedure for estimating RGP is as described by Stone and his co-workers and has not changed markedly over the past 25 years. It involves placing seedlings in a greenhouse or growth chamber which is programmed at conditions near optimum for root growth for the species in question. Following some standard time period (normally 4 weeks) seedlings are washed out of the medium and the new root growth which occurred during this period is quantified in some manner.

Basic Root Growth Test

The first step in testing RGP is to decide upon an appropriate sample size. This will depend on the nature of the seedling population in question and the reasons for making the measurements. Material propagated from wild seed will be highly variable and require fairly large samples. This is in contrast to half- or full-sib orchard material. If you are concerned only with a quick and-dirty evaluation of a batch of stock, 15 to 20 seedlings randomly selected (from different packing bags) may be sufficient. On the other hand, if your aim is to detect small differences in RGP among seed lots, certain nursery treatments, etc., then larger samples are required.

Also, since seedling size can affect RGP, it may be desirable either to (1) select seedlings of relatively uniform size for the test (which will bias the results) or

(2) collect data on seedling height, caliper and weight, along with RGP, and analyse the results using the morphological properties as covariates. Before potting, root systems are normally washed briefly to make them more visible and any white tips that might be present are pinched out. Then the seedlings are potted in large (15-L or 5-gal) plastic pots-normally 5 to 10 per pot depending upon size -and each pot is considered as one replicate.

As a potting mix, we have had excellent results with a 1:1 peat: vermiculite forestry mix, although many mixes have been used successfully. It is important that the mix be well drained, however, and sand or perlite can be incorporated for this purpose if necessary. Pots are watered initially to saturation and then maintained near field capacity by watering every two or three days. Fertiliser is not needed since the test will probably not go beyond 30 days. Seedlings have adequate nutrient reserves to carry them through this period. Some workers recommend liming the medium to maintain pH near 5.5, but we do not do this.

The key point to remember is that the test environment must remain constant from test to test. If this is not done, then differences in RGP between tests cannot be ascribed only to seedling condition. For maintaining constant environmental conditions, growth chambers are far superior to greenhouses. Realistically, however, few operational growers can afford growth chambers for seedling testing, while inexpensive greenhouses are within the means of many. Most RGP testing is therefore done in greenhouses, despite control problems.

A simple and easy-to-maintain environmental regime which promotes root growth in most northwest conifers is a constant 200 day/ night temperature and a 16-hour photoperiod, normally supplemented with fluorescent lights. It would be desirable to control relative humidity as well but this is rarely done due to its difficulty and expense. Since most RGP testing is done in winter and early spring, the greenhouse must be heated. If testing is done in late spring or summer -say, for high elevation stock-a shade cloth is essential for temperature control. Considerable errors can arise from extreme temperature fluctuations.

Quantifying Root Growth

Following the 4-week test period, seedlings are removed for root counting. This is easily done with a low pressure hose, using gentle force and care not to break the brittle new roots. If it is not possible to measure the roots immediately, they may be stored in a cold, dark room or refrigerator for several days before counting. This may be done either before or after unpotting. New roots can easily be distinguished from old roots by colour and general appearance. Old roots will have a rough surface and be dark brown. New ones will have pearly white tips and grade to a light tan and then brown as they become suberised. The point of transition from old to new root can usually be detected as a fairly abrupt change in diameter and/or colour.

Quantifying root growth can involve counting and/or measuring the length of new roots present. The intensity of measurement should depend upon test objectives. In some cases it may be sufficient to determine only whether or not any root growth has occurred. In other cases, you may need to count and measure each new root. An efficient method of root counting, is first to break the root system down into its major laterals, then count each lateral system individually. This can be done while holding the roots against a dark background and tallying with a lab counter.

Interpretation of the results of the RGP test depends on how the information will be used. First you should be generally familiar with the species you are working with and what sorts of RGP values are normal for that species. The test procedures described are not without problems and sources of error. First, it is easy to break and lose roots during the extraction process, especially if a heavy potting medium is used. This, obviously, will introduce measurement errors. Secondly, if environmental conditions change from test to test this will also introduce error.

Sources of these problems can involve changes in photoperiod, potting medium, soil moisture, soil and air temperature and other variables. Errors can also arise in the quantification procedure, especially if indirect methods, such as volume displacement, are used. It is also important to standardise measurement procedures to some extent. There can be considerable variation, for example, among root counters. So the same person or persons should be employed to count roots across treatments rather than having one person assess results of only one treatment.

Alternative Testing Methods

Aside from the various technical problems mentioned above, the major disadvantage of the RGP test is its cost. High cost results from (1) the time required to perform the test, during which seedlings must be watered and cared for, and (2) the labour-intensive process of quantifying new root growth. There have been numerous attempts to circumvent these problems. Burdett reported on an RGP test which requires only seven days to conduct. The main concept of this accelerated test is to speed up root growth by raising temperature. The conditions he used are described below.

Test condition

Day temperature	30°C
Night temperature	25°C
Daily photoperiod	16 hours
Light intensity	25,000 lux
Relative humidity	75%

Watering schedule

Start of test Ist watering	*2nd watering*	*3rd watering*
Monday a.m.	Wednesday noon	Friday a.m.
Wednesday noon	Friday p.m.	Monday a.m.
Friday a.m.	Monday a.m.	Wednesday noon

We tried the accelerated method with lodgepole pine and interior spruce (*Picea engelmannii glauca* complex) and were impressed with how well it worked. Burdett (pers. comm.) has also experimented with interior Douglas-fir (*Pseudotsuga menziesii glauca* (Mayr) Sudw) and coastal Douglas-fir. With both species he reports that results of the seven-day test are well correlated with results of 30-day tests. The only drawback of this test is that it requires a controlled environment chamber.

We have also had good results with a hydroponic test system. A 40-L (10 gallon) aquarium was painted black and covered with a black plywood lid. The water was aerated with an aquarium pump and bubble stone. No mineral nutrients were added. In a series of trials we measured RGP of Douglas-fir seedlings in the aerated water baths and compared results to RGP measured in the standard manner on subsamples of the same seedlings. Results of these trials showed good agreement between the two methods in terms of number and total length of roots produced per seedling.

The hydroponic approach has a number of advantages over the pot approach:

1) The time-consuming process of potting and unpotting seedlings is avoided as is the need for large quantities of expensive potting medium,
2) much greater spatial and temporal uniformity is achieved in the rooting environment,
3) assessment of new root growth is more accurate because (a) new roots are clean, not stained by the potting medium, and therefore easily distinguished from old roots, and (b) roots are not broken or lost in the unpotting process,
4) the technique facilitates alternative root counting methods, such as photo graphy and liquid displacement (discussed later), which are rapid and nondestructive,
5) new root growth can be monitored visually during the test, which may offer an opportunity for shortening the test period,
6) trials require about 50% less bench space,
7) water baths require no maintenance, while pots need frequent watering.

Mist or aeroponic systems may also offer an advantage over potting for clean, rapid root growth measurement. In some tests mortality and water stress were unexpectedly high for no apparent reason. New fogging nozzles and other aeroponic devices are now available, however, which might make this approach very attractive.

Alternative Methods for Quantifying Root Growth

Once the test has been completed, the next task is to determine how much root growth occurred. Much work has been done to simplify this tedious procedure. Initially, workers tended to count and measure total length of all new roots produced on each seedling. This apparently led to the realisation that the two values tend to be fairly well correlated. This gave rise to a subsequent approach of counting only. This alone halves the time involved. There have been other short cuts-counting only the tips longer than some critical length, measuring only the three or five longest new roots, tallying only the percent of seedlings tested which showed any new root growth at all, and etc. Each of these procedures might be appropriate under some circumstances but not others.

Burdett has developed an index which involves stratifying RGP results into six classes based upon numbers of new roots greater or less than some critical value:

Class	Description
0	No new root growth
1	Some new roots but none over 2 cm long
2	1-3 new roots over 1 cm long
3	4-10 new roots over 1 cm long
4	11-30 new roots over 1 cm long
5	More than 30 new roots over 1 cm long

This has proven to be a good predictor of field survival for some species. Another valid approach is to develop a set of reference photographs of seedlings with known RGP. The technician then matches a root system in hand with the photo which it most closely resembles.

Some workers clip off new roots and weigh them as an estimate of RGP. This method is as labourious as measuring. Still others use volumetric determinations. One approach involves placing a large beaker of water on a weighing balance. If a root system (or any object) is submerged in the water, the balance will record the weight of water displaced. This weight in grams is essentially equal to the volume of the root system in cm^3 (since one cm^3 of water weighs 1 g at STP). Volume determinations are made on a root system before and after a root growth test. These are then subtracted to give an estimate of the volume gained during the test. This then is a measure of RGP.

While elegant and simple, the method does suffer from two serious problems. First, if any part of the root system touches the sides or bottom of the beaker it will introduce an error in the volume determination. Second, if the same root system is measured several times in succession, different values are obtained each time. This may be due to absorption of water by the roots or incomplete drying of the roots between measurements or some other cause. For these reasons, we have been unable to obtain consistent, accurate RGP values using this technique.

The same problem with water absorption operates here also, however, and the success with this method has also been limited. This is not to discourage others from trying either method, however. Morrison and Armson reported on a device they developed and called a "rhizometer" for measuring root area and volume. It employs a photoelectric cell and provides rapid, nondestructive estimates of root area or volume.

While it appears to have some promise in RGP assessment, the method does not seem to have caught on. Root systems are photographed against grids and grid intersections counted and calibrated against total root length. The method is quite accurate but is more laborious than hand measuring.

Predictive Ability

A major objective of seedling quality assessment is to predict field survival and performance. Thus it is surprising that more data on the relationship between RGP and postplanting performance are not available. A cursory search of the English literature turned up about twenty such studies-most with conifers. In many cases, good agreement was reported between RGP and survival, both in field and greenhouse trials. In an early study, Stone found that RGP was a fairly good predictor of survival in jeffrey pine (Pinus *jeffreyi* Grev. and Bal.), ponderosa pine *ponderosa* Laws.), white fir (Abies *concolor* Gord. & Glend. Lindl.), red fir (A. *magnifica* A. Murr.) and Douglas-fir.

Seedlings of these species which failed to initiate roots within 60 days did not survive an additional 120 days in the greenhouse. Rhea found a close correlation between RGP and field survival in loblolly pine (*Pinus* taeda L.), as did Burdett et al. with white spruce (Picea *glauca* (Moench.) Voss.) and odgepole pine, and McMinn with lodgepole pine. Todd evaluated the predictive ability of RGP with stored and unstored Douglas-fir seedlings. With unstored seedlings survival was high and not well correlated with RGP. With stored seedlings, however, RGP accounted for 53% of the variability in survival. RGP and survival were clearly related in ponderosa pine and white fir seedlings. In pine, survival up to 95% was linear with RGP. White fir gave a similar response.

In some cases, RGP has also been a good predictor of growth after planting. Notable among these are the data of von Althen and Webb with sugar maple and Burdett et al. with lodgepole pine and white spruce. In the latter study, RGP accounted for 82% and 96% of the variability in first year shoot extension of pine and spruce, respectively. Webb found that the difficult-to regenerate sugar maple tended to have weaker RGP than the more easily regenerated white ash (Fraxinus *americana* L.) and silver maple (Acer *saccharinum* L.). He speculated that RGP might be an important factor in regeneration success. In contrast, there was no relationship between RGP and survival for sweetgum (*Liquidambar styraciflua* L.) and yellow poplar (Liriodendron tulipifera L.) and an inverse relation in sycamore (*Platanus occidentalis.*)

There are a few examples of studies with conifers in which RGP did not predict performance. In an extensive field trial with jack pine and black spruce, Sutton was not able to demonstrate this relationship, probably due to diverse weather and site conditions. He also found, in a nursery outplant trial with jack pine, that large RGP differences were not reflected in survival differences. With spruce, however, RGP differences of an order of magnitude were correlated with significant performance differences.

Interestingly, Sutton examined root growth of seedlings planted in the field and found no relationship with measured values of RGP before planting. Brissette and Roberts found only low correlations of RGP with survival and height growth in loblolly pine. They note, however, that survival was excellent (89%) and that first year field conditions were not very stressful. This may have obscured the sought-after relationships. So, although there is not total agreement on the ability of RGP to predict field performance, the weight of available evidence to date does indicate a strong relationship -at least with conifers in the Northwest.

8

Seed Health Testing Methods

Seed health testing deals with detection and identification of seed-borne microorganisms. The objective of seed health testing is to determine the health status of a seed sample, and by inference of the seed lot, thus gaining information that can be used to compare the value of different seed lots. Seed health testing may elucidate seedling evaluation and causes of poor germination or field establishment and thus supplement germination testing.

Seed testing is carried out to determine the purity (i.e. the percentage of pure seed, seeds of other crops and inert matter), the germination potential of a seed lot, etc.

There are various seed health testing methods:

Blotter Method

Blotter Method is used for the detection of *Alternaria dauci* on *Daucus carota*.

Materials

Reference material	-	The use of reference cultures or other appropriate material is recommended.

Substrate	-	Blotters or filter papers, 9.0 cm, circular (e.g. Whatman No 1 or equivalent), free from micro-organisms and inhibitors (3 per plate).
Plates	-	9.0 cm sterile
Petri dishes		(one per ten seeds).
Incubator	-	Operating at 20 ± 2°C, equipped with timer-controlled nearultraviolet lights (NUV, peak at 360 nm, e.g. colour number 08, Philips; BLB, Sylvania).
Freezer	-	Operating at -20 ± 2°C.

Seed Sample Preparation

1. It is vital to exclude any possibility of cross-contamination between seed samples.

   This can be achieved by swabbing/spraying equipment and gloved hands with 70% ethanol.

Method

1. Place three 9.0 cm filter papers in each plate and soak with sterile distilled/de-ionised water. Drain away excess water.
2. Aseptically place 10 seeds, evenly spaced (CCP), on the surface of the filter paper in each plate.
3. Incubate for 3 d at 20 ± 2°C in the dark.
4. Transfer plates to freezer and maintain at -20 ± 2°C for 24 h.
5. After freezing, incubate for 6 d at 20 ± 2°C with alternating 12 h periods of darkness and NUV light.

Plates should be approx. 25 cm below the lights and should not be stacked.

6. Examine seeds under a stereoscopic microscope at x30 for fungal growth and up to x80 magnification for identification of conidia. Conidiophores are simple or slightly branched, arising singly or in small groups from the surface of the seed or on aerial mycelium. Conidia are usually solitary, obclavate, up to 450 µm long, pale olivaceous brown at first, becoming brown with age, with a long pale beak up to 3 times the length of the body. Groups of sunken conidia are sometimes visible by the emerging clusters of their bright long beaks. Record the number of infected seeds in each plate (CCP).

General Methods

Checking tolerances

Tolerances provide a means of assessing whether or not the variation in result within or between tests is sufficiently wide as to raise doubts about the accuracy of the results.

Reporting results

The result of a seed health test should indicate the scientific name of the pathogen, and the test method used. When reported on an ISTA Certificate, results are entered under other determinations. In the case of a negative result (pathogen not detected), the results should be reported in terms of the tolerance standard (e.g. infection level less than 1% with 95% probability). The tolerance standard depends on the total number of seeds tested, n, and is approximately 3/n (P=0.95). In the case of a positive result the report should indicate percentage of infected seeds.

Quality Assurance

Specific training

This test should only be performed by persons who have been trained in fungal identification or under the direct supervision of someone who has.

Critical control points

Spreading hyphae may lead to contamination of other seeds. Seeds must therefore be spaced at least 20 mm from each other with not more than 10 seeds per 9.0 cm Petri dish.

Samples may be difficult to examine due to the growth of contaminants, especially Alternaria tenuis, and/or A. radicina. Experience and great care is required for the detection of all occurrences.

Supplementary examination may be required 13-14 d after plating of seeds to allow hampered or deeper inoculum to exceed contaminants by its fructification.

Adequate Safety Precautions

Ensure you are familiar with hazard data and take appropriate safety precautions. It is assumed that this procedure is being carried out in a microbiological laboratory by persons familiar with the principles of Good Laboratory Practice, Good Microbiological Practice, and aseptic technique. Dispose of all waste materials in an appropriate way (e.g. autoclave, disinfect) and in accordance with local health, environmental and safety regulations.

Treated seed

Seed treatments may affect the performance of this test. It should only be performed on untreated seed.

Malt Agar Method

Malt Agar Method is used for the detection of *Alternaria dauci* on *Daucus carota*

Materials

Reference material	-	The use of reference cultures or other appropriate material is recommended.
Malt Agar plates	-	6. 9.0 cm plates (Petri dishes; one plate per ten seeds).
Incubator	-	Operating at 20 ± 2°C, equipped with timer-controlled nearultra-violet lights.

Sample Preparation

1. It is vital to exclude any possibility of cross-contamination between seed samples.

   This can be achieved by swabbing/spraying equipment and gloved hands with 70% ethanol.

Method

1. Aseptically place a maximum of 10 seeds evenly spaced on the agar surface of each malt agar plate.
2. Incubate plates for 10 d at 20 ± 2°C, with alternating 12 h periods of darkness and NUV light. Plates should be approx. 25 cm below the lights and should not be stacked.
3. Sub-culture a reference culture to a malt agar plate at the same time seeds are plated and incubate with the test plates.

4. Examine plates visually, and under a stereoscopic microscope at x30 magnification, for fungal growth. Use a magnification of x50 - x80 for identification of conidia. Colonies of Alternaria dauci are brown or dark brown with olive-grey aerial mycelium and produce a brown diffusible pigment in the medium. Conidiophores are simple or slightly branched, arising singly or in small groups from the surface of the seed or on aerial mycelium. Conidia are usually solitary, obclavate, up to 450 μm long, pale olivaceous brown at first, becoming brown with age, with a long pale beak up to 3 times the length of the body. Groups of sunken conidia are sometimes visible by the emerging clusters of their bright long beaks. Record the number of infected seeds in each plate (CCP).

Common Methods

1. Checking tolerances

Tolerances provide a means of assessing whether or not the variation in result within or between tests is sufficiently wide as to raise doubts about the accuracy of the results.

2. Reporting results

The result of a seed health test should indicate the scientific name of the pathogen and the test method used. When reported on an ISTA Certificate, results are entered under other determinations. In the case of a negative result (pathogen not detected), the results should be reported in terms of the tolerance standard (e.g. infection level less than 1% with 95% probability). The tolerance standard depends on the total number of seeds tested, n, and is approximately 3/n (P=0.95).

In the case of a positive result the report should indicate percentage of infected seeds.

Quality Assurance

Specific Training

This test should only be performed by persons who have been trained in fungal identification or under their direct supervision.

Critical Control Points

Contaminants may greatly compete with the pathogen on the non-selective medium making detection laborious and difficult. The malt agar source can influence the results. Whenever a new batch of malt agar is used a check on the quality should be made using a reference lot with a known infection level.

Preparation of Malt Agar

Compound	*g/l*	*g/500ml*
Malt Agar (CCP)	as specified by manufacturer	
De-ionised/distilled Water	1000 ml	500 ml

Preparation

1. Weigh out ingredients into a suitable autoclavable container.
2. Add 1000 (or 500) ml of distilled water.
3. Steam to dissolve.
4. Autoclave at 15 PSI and 121°C for 15 min.
5. Allow agar to cool to approx. 50°C.
6. Pour 15-22 ml of molten agar into 9.0 cm Petri plates and allow to solidify at room temperature (20-25°C) for 24 h before use.

Storage

Prepared plates may be stored at room temperature or at 4°C for up to one month before use.

Safety Precautions

Ensure you are familiar with hazard data and take appropriate safety precautions, especially during preparation of media, autoclaving, and weighing out of ingredients. It is assumed that this procedure is being carried out in a microbiological laboratory by persons familiar with the principles of Good Laboratory Practice, Good Microbiological Practice, and aseptic technique. Dispose of all waste materials in an appropriate way (e.g. autoclave, disinfect) and in accordance with local health, environmental and safety regulations.

Treated seed

Seed treatments may affect the performance of this test. It should only be performed on untreated seed.

Botrytis Cinerea

Botrytis cinerea is used the detection of *Helianthus annuus* (Sunflower)

Materials

Reference Material	-	The use of reference cultures or other appropriate material is recommended when ever possible.
Media	-	Blotters (Whatman No 1) 3% malt solution
Petri dishes	-	When sowing density is given by a number of seeds per petri dish, a diameter of 90 mm is assumed.

Incubator - Capable of operating in the range 20°C ± 2°C.

Method

1. *Pretreatment*

   None.

2. *Blotter*

Place two pieces (88 mm in diameter) of filter paper in each of 80, 90 mm diameter Petri dish, and add 5 ml of a solution of 3% malt extract. Pour off the excess liquid and plant 5 seeds in each petri dish.

3. *Incubation*

   9 days at 20°C in darkness.

4. *Examination:* After 5, 7 and 9 days examine the seeds by naked eye for roots showing a soft rot and covered by an abundant grey mycelium. These seeds are recorded as infected. Examination is repeated after 7 and 9 days incubation. New infections are marked on the blotter, and the sum of the three counts gives the total infection. In doubtful cases confirmation may be made by examining the mycelium at x200 magnification for septae, ribbon-like hyphae and tufts of branching conidiophores.

Common Methods

Checking tolerances

Tolerances provide a means of assessing whether or not the variation in result within or between tests is sufficiently wide as to raise doubts about the accuracy of the results.

Reporting results

The result of a seed health test should indicate the scientific name of the pathogen detected and the percentage of infected seeds. When reported on an ISTA Certificate results are entered under Other Determinations. The results should be accompanied by information on the test method used, including any pretreatment.

Safety Precautions

Ensure you are familiar with hazard data and take appropriate safety precautions, especially during preparation of media, autoclaving and weighing out of ingredients. Dispose of all waste materials in an appropriate way (e.g. autoclave, disinfect) and in accordance with local safety regulations.

Treated seed

This method may be influenced by treatment applied to the seed lot. Seed health tests on chemically treated seeds will generally deliver unreliable test results caused by masking or inhibition of the growth of the target organism.

Phoma lingam

Phoma lingam method is used the detection on *Brassica spp.*

Materials

Reference Material - The use of reference cultures or other appropriate material is recommended when ever possible.

Media	- Blotters
2,4 - Dichlorophenoxyacetate (2,4D)	- Sodium salt 0.2% solution
Petri dishes	- When sowing density is given by a number of seeds per petri dish, a diameter of 90 mm is assumed.
Incubator	- Capable of operating in the range 20ºC ± 2ºC. To stimulate sporulation, alternating 12-hour periods of darkness and nearultra-violet light (NUV) during incubation are recommended. The recommended source is the black light fluorescent lamp (peak at 360 nm) but daylight fluorescent tubes are satisfactory.

Sample Preparation

The test is carried out on a working sample of 1000 pure seeds.

Method

1. Blotter

Place three pieces of filter paper in each petri dish and add 5 ml of a 0.2% solution of the sodium salt of 2,4 dichlorophenoxyacetic acid (2,4-D) to inhibit seed germination. Pour of the excess 2,4-D solution, rinse the seeds in sterile water and place 50 in each dish.

2. *Pretreatment*

   None.

3. Incubation

   Incubate for 11 days at 20°C with alternating cycles of 12 hours light and 12 hours darkness.

4. *Examination*

   After 6 days examine at x25 magnification for loose growing silver white mycelium and pycnidial primordia of Phoma lingam on the seed and substrate. After 11 days make a second examination for pycnidia on infected seeds and on the filter paper near infected seeds. Seeds from which pycnidia of Phoma lingam have developed are recorded as infected.

   Pycnidia are relatively large, about 250μ, with papilla, some times developed as a neck. The ubiquitous sapro phyte Phoma herbarum Westend occurs also on brassica seed, but has smaller pycnidia formed superficially on the seed coat, not papillate, with white yellow or pink but not purple (amethyst) exudate.

Common Methods

Checking tolerances

Tolerances provide a means of assessing whether or not the variation in result within or between tests is sufficiently wide as to raise doubts about the accuracy of the results for most direct seed health tests.

Reporting results

The result of a seed health test should indicate the scientific name of the pathogen detected and the percentage of infected seeds. When reported on an ISTA Certificate results are entered under Other Determinations. The results should be accompanied by information on the test method used, including any pretreatment.

Safety Precautions

Ensure you are familiar with hazard data and take appropriate safety precautions, especially during preparation of media, autoclaving and weighing out of ingredients.

Treated Seed

This method may be influenced by treatment applied to the seed lot. Seed health tests on chemically treated seeds will generally deliver unreliable test results caused by masking or inhibition of the growth of the target organism.

Ascochyta Pisi

Ascochyta pisi method is used to the detection on *Pisum sativum* (Pea)

Materials

Reference Material	-	The use of reference cultures or other appropriate material is recommended when ever possible.
Media	-	Malt Agar or Potato Dextrose Agar.
Sodium hypochlorite solution	-	(1% available chorine) for seed disinfection.
Petri dishes	-	When sowing density is given by

		a number of seeds per petri dish, a diameter of 90 mm is assumed.
Incubator	-	Capable of operating in the range 20°C ± 2°C.

Method

1. *Pretreatment*

   10 minutes in a 1% (available chlorine) sodium hypochlorite solution followed by draining.

2. *Agar Method*

   Malt or potato dextrose agar. Place 10 seeds on the agar surface in each petri dish.

3. *Incubation*

   7 days at 20°C in darkness.

4. *Examination*

   After 7 days examine each seed by naked eye for abundant white mycelium which often covers infected seeds. Doubtful colonies may be confirmed by the presence of wavy hyphae at the edge of the colony when examined at x25 magnification. Colony diameter typically 20-30 mm, occasionally smaller or incompletely surrounding the seed. Reverse of colonies medium to dark orange-brown centrally, opaque and even, paleing to a narrow creamy border. Gelatinous-looking or ange-brown pycnidia often present (although only sometimes clearly visible), particularly where seed touches agar. Under STM at x20-25 magnification, using both transmitted and incident light, hyphae are curled, often several running together, typically with moisture drops (although these evaporate easily). Very limited growths from some seeds may only be seen if dishes tilted to get lighting at best angle, or under STM examination or after extended incubation. Pycnidia are up to 250μ a diameter. Spores, hyaline, cylindric, of slightly curved with rounded ends, 1-septate, slightly constricted at septum, mostly 12μ x 4.5μ.

Common Methods

Checking tolerances

Tolerances provide a means of assessing whether or not the variation in result within or between tests is sufficiently wide as to raise doubts about the accuracy of the results.

Reporting results

The result of a seed health test should indicate the scientific name of the pathogen detected and the percentage of infected seeds.

Preparation of Media and Solutions

1. Sodium Hypochlorite Solution

Sodium hypochlorite for pretreatment of seed can be pre pared from commercial bleach diluted to 1% available chlorine. The concentration of chlorine in commercial bleach varies considerably. Use the formula

$$V_{stock} = V_{final} \times C_{final} / C_{stock}$$

(where V= volume and C= % available chlorine) to calculate the volume of commercial bleach stock solution required to prepare sodium hypochlorite solutions for use in seed pre treatment.

To prepare a 1 liter solution of sodium hypochlorite containing 1% chlorine from a stock of commercial bleach containing 12% available chlorine:

$V_{stock} = V_{final} \times C_{final} / C_{stock}$ $V_{stock} = 1L \times 1/12 = 0.083$

Thus add 83 ml of the 12% stock to 917 ml water.

2. Malt Agar

Compound	g/l	g/500ml
Malt Agar	45	22.5
Deionised/distilled Water	1 l	500 ml

Preparation

1. Weigh out ingredients into a suitable autoclavable container.
2. Add 1000 (or 500) ml of deionised/distilled water.
3. Dissolve powdered Malt Agar in deionised/ distilled water by stirring.
4. Autoclave at 15 PSI and 121°C for 15 min.
5. Allow agar to cool to approx. 50°C.
6. Pour 15-22 ml of molten agar into 90 mm Petri plates and allow to solidify before use.

   Plates may be stored at room temperature in a dry cabinet or at 4°C.

3. Potato Dextrose Agar

Compound	g/l	g/500ml
Potato Dextrose Agar	39	19.5
Deionised/distilled Water	1 l	500 ml

Preparation

1. Weigh out ingredients into a suitable autoclavable container.

2. Add 1000 (or 500) ml of deionised/distilled water.
3. Dissolve powdered PDA in deionised/distilled water by stir ring.
4. Autoclave at 15 PSI and 121°C for 15 min.
5. Allow agar to cool to approx. 50°C.
6. Pour 15-22 ml of molten agar into 90 mm Petri plates and allow to solidify before use.

   Plates may be stored at room temperature in a dry cabinet or at 4°C.

Safety Precautions

Ensure you are familiar with hazard data and take appropriate safety precautions, especially during preparation of media, autoclaving and weighing out of ingredients. Dispose of all waste materials in an appropriate way (e.g. autoclave, disinfect) and in accordance with local safety regulations.

Treated Seed

This method may be influenced by treatment applied to the seed lot. Seed health tests on chemically treated seeds will generally deliver unreliable test results caused by masking or inhibition of the growth of the target organism.

Colletotrichum Lindemuthianum

Colletotrichum lindemuthianum is used to the detection on *Phaseolus vulgaris* (Bean)

Materials

Reference Material - The use of reference cultures or other appropriate material is recommended when ever possible.

Media	-	Paper towelling.
Sodium hypochlorite solution	-	(1% available chorine) for seed disinfection.
Incubator	-	Capable of operating in the range 20°C ± 2°C.

Sample Preparation

The test is carried out on a working sample of 400 seeds.

Method

1. *Pretreatment*

   Seeds are submerged in a solution of 1% (available chlorine) sodium hypochlorite for 10 minutes and allowed to drain.

2. *Between Paper (BP)*

   Spread the seeds in replicates of 50 on double sheets of paper towelling 350 x 450 mm which have been soaked in water. Cover seeds with one sheet of paper towelling soaked in water. Fold the paper twice lengthways and cover it with a sheet of polythene to maintain the moisture during incubation.

3. *Incubation*

   7 days at 20°C in darkness

4. *Examination*

   After 7 days remove the seed coats and examine the cotyledons by naked eye for black depressed areas with well de limited out lines. Check each spot for the presence of acervuli with or without dark brown setae using x25 magnification. The septate setae measure approx. 6μ x 100μ. The pale orange acervuli contain cylindric, hyaline conidia with rounded ends containing one or two guttulaes. Conidia measure 2.5 5.5μ x 11 20μ. The use of a high power microscope (magnification x200) is sometimes necessary. Small spots may require longer incubation for development of acervuli.

Common Methods

Checking tolerances

Tolerances provide a means of assessing whether or not the variation in result within or between tests is sufficiently wide as to raise doubts about the accuracy of the results.

Reporting results

The result of a seed health test should indicate the scientific name of the pathogen detected and the percentage of infected seeds. The results should be accompanied by information on the test method used, including any pretreatment.

Preparation of Media and Solutions

1. Sodium Hypochlorite Solution

Sodium hypochlorite for pretreatment of seed can be prepared from commercial bleach diluted to 1% available chlorine. The concentration of chlorine in commercial bleach varies considerably. Use the formula $V_{stock} = V_{final} \times C_{final} / C_{stock}$ (where V = volume and C = % available chlorine) to calculate the volume of commercial bleach stock solution required to prepare sodium hypochlorite solutions for use in seed pretreatment. To prepare a 1 liter solution of sodium hypochlorite containing 1% chlorine from a stock of commercial bleach containing 12% available chlorine:

$V_{stock} = V_{final} \times C_{final} / C_{stock}$ $V_{stock} = 1L \times 1/12 = 0.083$

Thus add 83 ml of the 12% stock to 917 ml water.

Safety Precautions

Ensure you are familiar with hazard data and take appropriate safety precautions, especially during preparation of media, autoclaving and weighing out of ingredients. It is assumed that this procedure is being carried out in microbiological laboratory by persons familiar with the principles of Good Laboratory Practice, Good Microbiological Practice, and aseptic technique. Dispose of all waste materials in an appropriate way (e.g. autoclave, disinfect) and in accordance with local safety regulations.

Treated seed

This method may be influenced by treatment applied to the seed lot. Seed health tests on chemically treated seeds will generally deliver unreliable test results caused by masking or inhibition of the growth of the target organism.

Botrytis Cinerea

Botrytis cinerea is used to the detection on *Linum usitatissimum* (Flax)

Materials

Reference Material	-	The use of reference cultures or other appropriate material is recommended when ever possible.
Media	-	Malt Agar

Petri dishes	-	When sowing density is given by a number of seeds per petri dish, a diameter of 90 mm is assumed.
Incubator	-	Capable of operating in the range 20°C ± 2°C.

Method

1. *Pretreatment*

   None

2. *Agar Plate Method*

   Malt agar containing 2% agar and 1% malt extract.

3. *Sowing density*

   Place 10 seeds on the agar surface in each petri dish.

4. *Incubation*

   7 days at 20°C in darkness

5. *Examination*

   After 5 and 7 days examine for roots showing a soft rot and covered by abundant grey mycelium. Colonies on agar measure up to 5 cm in diameter after 5 days. Identification can be checked by high power microscope (magnification x200). Mycelium of tape-like hyphae producing bunches of branching conidiophores with ovoid-hyaline one-celled conidia 8 x 11 - 6 x 19μ. When analysts are familiar with the fungus, naked eye examination is sufficient for identification.

Common Methods

Checking tolerances

Tolerances provide a means of assessing whether or not

the variation in result within or between tests is sufficiently wide as to raise doubts about the accuracy of the results.

Reporting results

The result of a seed health test should indicate the scientific name of the pathogen detected and the percentage of infected seeds. The results should be accompanied by information on the test method used, including any pretreatment.

Preparation of Media

Malt Extract Agar

Compound	g/l	g/500 ml
Malt Extract	10	5
Agar (equivalent to oxoid No.3)	20	10
Deionised/Distilled Water	1 l	500 ml

Preparation

1. Weigh out ingredients into a suitable autoclavable container.
2. Add 1000 (or 500) ml of deionised/distilled water.
3. Dissolve powdered Malt Agar in deionised/distilled water by stirring.
4. Autoclave at 15 PSI and 121°C for 15 min.
5. Allow agar to cool to approx. 50°C.
6. Pour 15-22 ml of molten agar into 90 mm Petri plates and allow to solidify at room temperature before use.

Plates may be stored at room temperature in a dry cabinet or at 4°C.

Safety Precautions

Ensure you are familiar with hazard data and take appropriate safety precautions, especially during preparation of media, autoclaving and weighing out of ingredients. Dispose of all waste materials in an appropriate way (e.g. autoclave, disinfect) and in accordance with local safety regulations.

Treated Seed

This method may be influenced by treatment applied to the seed lot. Seed health tests on chemically treated seeds will generally deliver unreliable test results caused by masking or inhibition of the growth of the target organism.

Caloscypha Fulgens

Caloscypha fulgens is used to the detection on *Picea engelmannii* and *glauca* (Spruce)

Materials

Reference Material - The use of reference cultures or other appropriate material is recommended when ever possible.

Media - Water Agar.

Hydrogen Peroxide - (30%) for seed disinfection.

Petri dishes - When sowing density is given by a number of seeds per petri dish, a diameter of 90 mm is assumed.

Incubator - capable of operating in the range 15°C ± 2°C.

Method

1. Pretreatment

   Surface sterilise the seeds for 30 min. in 30% hydrogen per oxide (three volumes of the H_2O_2 per one volume of seeds).

   CCP The 30 min. surface sterilisation with 30% H_2O_2 signicantly reduces contamination from other fungi and bacteria and allows better detection of *C. fulgens*; surface sterilisation for periods longer than 30 minutes decreases incidence of the pathogen.

   Stir the seeds once or twice during the 30 minutes.

   Drain of the hydrogen peroxide with and agitate the seeds for 5 minutes in sterile, distilled water; then drain off.

   Surface dry the seeds on sterile paper in a sterile environment.

2. *Medium Water Agar*

   Place the surface sterilised seed onto 1.5% water agar, about 15 ml/9 cm petri dish. 25 seeds are placed in a petri dish.

3. *Incubation*

   Store the plates in plastic bags at 15°C for 5 weeks under fluorescent light for 8-12 hour intervals of alternation with darkness.

4. *Examination*

   Examine every 3 days and remove seeds exhibiting blue stain (produced by *C. fulgens*) in agar or typical coarse varrucose, right angle branched hyphae usually covered with water droplets. Characteristics are identiable at x100 using a stereomicroscope. Seeds that germinate do not yield the fungus. The conidiophores which arise from aerial hyphae are 200- 550μ high, smooth pale yellow to yellow brown below, 8-17μ in diameter below and taper to 3.6 6μ above; unbranched up to 220μ, then more or less dichotomously branched; hyaline above. Conidia are 4.6-7.6μ x 3.2-4.0μ; holoblastic, smooth, hyaline, ovate, dry. More details on the character is tics of the hyphae, conidiophores and conidia of *C. fulgens* are given by Paden et al. and Salt.

Common Methods

Checking tolerances

Tolerances provide a means of assessing whether or not the variation in result within or between tests is sufficiently wide as to raise doubts about the accuracy of the results.

Reporting results

The result of a seed health test should indicate the scientific name of the pathogen detected and the percentage of infected seeds.

Preparation of Media and Solutions

1.5% Water Agar

Compound	g/l	g/500 ml
Agar	15	7.5
Deionised/Distilled Water	1 1	500 ml

Preparation

1. Weigh out ingredients into a suitable autoclavable container.

2. Add 1000 (or 500) ml of deionised/distilled water.
3. Dissolve powdered Agar in deionised/distilled water by stirring.
4. Autoclave at 15 PSI and 121°C for 15 min.
5. Allow agar to cool to approx. 50°C.
6. Pour 15ml of molten agar into 90 mm Petri plates and allow to solidify at room temperature before use.

   Plates may be stored at room temperature in a dry cabinet or at 4°C.

Safety Precautions

Ensure you are familiar with hazard data and take appropriate safety precautions, especially during preparation and handling of hydrogen peroxide (a strong oxidising agent), media, autoclaving and weighing out of ingredients. It is assumed that this procedure is being carried out in microbiological laboratory by persons familiar with the principles of Good Laboratory Practice, Good Microbiological Practice, and aseptic technique. Dispose of all waste materials in an appropriate way (e.g. autoclave, disinfect) and in accordance with local safety regulations.

Fusarium moniliforme var. *subglutinan*

Fusarium moniliforme var. *subglutinan* is used to the detection on *Pinus taeda* and *P. elliotii* (Pine)

Materials

Reference Material - The use of reference cultures or other appropriate material is recommended when ever possible.

Media	-	Blotters
PCNB	-	Liquid solution.
Plastic containers	-	133 x 133 x 32 mm.
Incubator	-	Capable of operating in the range 20°C ± 2°C. The ability to alternate fluorescent light and darkness during incubation is required.

Method

1. *Pretreatment*

   None.

2. *Blotter Method*

   PCNB liquid solution.

3. *Plating*

   On blue blotter paper in plastic containers (133 mm x 133 mm x 32 mm) place 25 seeds evenly spaced in each container and crush with a sterilised piece of plastic designed to fit the container. Spray seed and blotter with PCNB liquid solution. Cover with transparent lid.

4. *Incubation*

   10-16 days at 20°C under fluorescent light at convenient intervals of alternation with darkness until the colonies are about 2 cm in diameter.

5. *Examination*

   Examine each colony at a magnification of x100-400 for the presence of and appearance of polyphialides and conidia. A colony is classified as being of this species if characteristic polyphialides, microconidia, and macroconidia are present. Do not include colonies with microconidia in chains, pear shaped microconidia, or with chlamydospores even if polyphialides are present.

Common Methods

Checking tolerances

Tolerances provide a means of assessing whether or not the variation in result within or between tests is sufficiently wide as to raise doubts about the accuracy of the results.

Reporting results

The result of a seed health test should indicate the scientific name of the pathogen detected and the percentage of infected seeds. The results should be accompanied by information on the test method used, including any pretreatment.

Media and Solutions

PCNB Liquid Solution.

Compound	g/l	g/500 ml
Peptone	15	7.5
$MgSO_4 \cdot 7H_2O$	5	2.5
KH_2PO_4	1	0.5
75% wettable powder (PCNB) (Terraclor 75% wettable powder)	1	0.5
Deionised/Distilled Water	1 l	500 ml
Streptomycin Sulphate	1	0.5
Neomycin Sulfate	0.12	0.06

Preparation

1. Weigh out ingredients (except streptomycin sulphate and neomycin sulfate) into a suitable autoclavable container.
2. Add 1000 (or 500) ml of distilled water.
3. Dissolve powdered agar, peptone, $MgSO_4 \cdot 7H_2O$, KH_2PO_4 and PCNB in deionised H_2O by stirring.
4. Autoclave at 15 PSI and 121°C for 15 min.
5. Cool to room temperature.
6. Add streptomycin sulfate and neomycin sulfate.
7. Use as described in method.

Safety Precautions

Ensure you are familiar with hazard data and take appropriate safety precautions, especially during preparation of media, PCNB solutions, autoclaving and weighing out of ingredients. Dispose of all waste materials in an appropriate way (e.g. autoclave, disinfect) and in accordance with local safety regulations.

Drechslera Oryzae

Drechslera oryzae is used to the detection on *Oryza sativa* (Rice).

Materials

Reference Material	-	The use of reference cultures or other appropriate material is recommended when ever possible.
Media	-	Blotters
Petri dishes	-	When sowing density is given by a number of seeds per petri dish, a diameter of 90 mm is assumed.

Incubator - Capable of operating in the range 22°C ± 2°C. To stimulate sporulation, alternating 12-hour periods of darkness and near-ultraviolet light (NUV) during incubation are recommended. The recommended source is the black light fluorescent lamp (peak at 360 nm) but daylight fluorescent tubes are satisfactory.

Method

1. *Pretreatment*

   None

2. *Blotter*

   On water-soaked blotters in petri dishes. Place 25 seeds in each dish.

3. *Incubation*

   7 days at 22°C in Near Ultra Violet light in 12 hr. light/12 hr. dark cycle.

4. *Examination*

   Examine each seed at 12-50 magnification for conidia of *D. oryzae*. Conidiophores of the fungus are produced on the seed coat and also on light grey aerial mycelium covering whole or part of the seed, giving a fluffy appearance. The fungus may occasionally spread on to the blotters. In doubtful cases confirmation may be made by examining conidia at x200 magnification. Conidia are crescent-shaped 35-107μ x 11-17μ light brown to brown, widest in the middle or below the middle and tapering to rounded ends.

Common Methods

Checking tolerances

Tolerances provide a means of assessing whether or not the variation in result within or between tests is sufficiently wide as to raise doubts about the accuracy of the results.

Reporting results

The result of a seed health test should indicate the scientific name of the pathogen detected and the percentage of infected seeds.

Safety Precautions

Ensure you are familiar with hazard data and take appropriate safety precautions, especially during preparation of media, autoclaving and weighing out of ingredients. Dispose of all waste materials in an appropriate way (e.g. autoclave, disinfect) and in accordance with local safety regulations.

Treated Seed

This method may be influenced by treatment applied to the seed lot. Seed health tests on chemically treated seeds will generally deliver unreliable test results caused by masking or inhibition of the growth of the target organism.

Pyricularia Oryzae

Pyricularia oryzae is used to the detection on *Oryza sativa* (Rice)

Materials

Reference Material - The use of reference cultures or

		other appropriate material is recommended when ever possible.
Media	-	Blotters
Petri dishes	-	When sowing density is given by a number of seeds per petri dish, a diameter of 90 mm is assumed.
Incubator	-	Capable of operating in the range 22°C ± 2°C. To stimulate sporulation, alternating 12-hour periods of darkness and near-ultraviolet light (NUV) during incubation are recommended. The recommended source is the black light fluorescent lamp (peak at 360 nm) but daylight fluorescent tubes are satisfactory.

Method

1. *Pretreatment*

   None.

2. *Blotter Method*

   On water soaked blotters in petri dishes. Place 25 seeds in each dish.

3. *Incubation*

   7 days at 22°C in alternating cycles of 12 hours darkness and 12 hours light, preferably NUV light.

4. *Examination*

   Examine each seed at x12-50 magnification for conidia of *P. oryzae.* Generally, the fungus produces small, inconspicuous, grey to green colonies on glumes, consisting of short, delicate, conidiophores carrying clusters of conidia at their tips. The growth rarely covers the whole seed. In doubtful cases confirmation may be made by examining conidia at x200 magnification. Conidia are typically obpyriform, hyaline, truncated with a short tooth at the base, 2-septate, usually with a pointed acute apex, 20-25μ x 9-12μ.

Common Methods

Checking tolerances

Tolerances provide a means of assessing whether or not the variation in result within or between tests is sufficiently wide as to raise doubts about the accuracy of the results.

Reporting results

The result of a seed health test should indicate the scientific name of the pathogen detected and the percentage of infected seeds. When reported on an ISTA Certificate results are entered under Other Determinations. The results should be accompanied by information on the test method used, including any pretreatment.

Alternaria Padwickii

Alternaria padwickii is used to the detection on *Oryza sativa* (Rice)

Safety Precautions

Ensure you are familiar with hazard data and take appropriate safety precautions, especially during preparation of media, autoclaving and weighing out of ingredients. Dispose of all waste materials in an appropriate way (e.g. autoclave, disinfect) and in accordance with local safety regulations.

Treated Seed

This method may be influenced by treatment applied to the seed lot. Seed health tests on chemically treated seeds will generally deliver unreliable test results caused by masking or inhibition of the growth of the target organism.

Materials

Reference Material	-	The use of reference cultures or other appropriate material is recommended when ever possible.
Media	-	Blotters
Petri dishes	-	When sowing density is given by a number of seeds per petri dish, a diameter of 90 mm is assumed.
Incubator	-	capable of operating in the range 22°C ± 2°C. To stimulate sporulation, alternating 12-hour periods of darkness and near-ultraviolet light (NUV) during incubation are recommended. The recommended source is the black light fluorescent lamp (peak at 360 nm) but daylight fluorescent tubes are satisfactory.

Method

1. *Pretreatment*

   None

2. *Blotter*

On water-soaked blotters in petri dishes. Place 25 seeds in each dish.

3. *Incubation*

   7 days at 22°C preferably in Near Ultra Violet light in 12 hr.light/12 hr. dark cycle.

4. *Examination*

   Examine each seed at x12-50 magnification for conidia of A. padwickii. Conidia are fusiform, first sub-hyaline, later straw coloured to golden brown with long beaks, borne singly either ion short conidiophores arising straight from the seed coat or on conidiophores within white-grey, fluffy, aerial mycelium. In doubtful cases confirmation may be made by examining conidia at x200 magnification. Conidia are 3-5 septate, often con stricted at the septa with characteristic conical basal cells and long beaks, 95-170μ x 11-20μ. Infected seeds or seedlings are often surrounded by a characteristic pink stain which diffuses out into the blotter and becomes more intense after 7 days.

Common Methods

Checking tolerances

Tolerances provide a means of assessing whether or not the variation in result within or between tests is sufficiently wide as to raise doubts about the accuracy of the results.

Reporting results

The result of a seed health test should indicate the scientific name of the pathogen detected and the percentage of infected seeds. The results should be accompanied by information on the test method used, including any pretreatment.

Safety Precautions

Ensure you are familiar with hazard data and take appropriate safety precautions, especially during preparation of media, autoclaving and weighing out of ingredients. Dispose of all waste materials in an appropriate way (e.g. autoclave, disinfect) and in accordance with local safety regulations.

Treated Seed

This method may be influenced by treatment applied to the seed lot. Seed health tests on chemically treated seeds will generally deliver unreliable test results caused by masking or inhibition of the growth of the target organism.

Ustilago Nuda

Ustilago nuda is used to the detection on *Hordeum vulgare* (Barley)

Materials

Reference material	-	Seed known to be infected or other appropriate material.
Incubator	-	Operating at 20 ± 2°C.
Brass sieves	-	1mm mesh, 3 required.
Microscope	-	with sub-stage illumination, x25 and x50 magnification.

Method

1. Embryo Method Working sample
   -- Two replicates of 100-120 g containing, depending on 1000 seed weight, 2000-4000 seeds.

2. Extraction and clearing of embryos

   – Place the working sample in one litre of a freshly prepared 5% aqueous solution of sodium hydroxide (NaOH) and maintain at 20°C for 24 hours.

CCP A weaker solution of NaOH or a lower temperature makes extraction difficult.

   – After soaking, the entire sample should be transferred to a suitable container and washed in warm water to separate the embryos, which appear through the softened pericarps.

   – Collect the embryos in a sieve of 1 mm mesh. Additional sieves of larger mesh can be used to collect pieces of endosperm and chaff.

   – Transfer the embryos to a mixture of equal quantities of glycerol and water in which further separation of the embryos and chaff can be made.

   – Transfer the embryos to a beaker containing 75 ml of fresh water-free lactophenol and clear them by maintaining the lactophenol at boiling point for approximately 30 seconds in a fume cupboard.

   – Transfer the embryos to fresh, slightly warm glycerol for examination.

3. Examination

   – Examine 1000 embryos from each replicate at x16-25 magnification with adequate substage illumination for the characteristic golden brown mycelium of *Ustilago nuda*.

   – Mycelium is approximately 3μ thick, is golden brown in colour and visible without a stain.

Infection may vary from a few strands of short hyphae to complete invasion of the scutellum tissues. Occasionally fungi other than Ustilago nuda occur in the scutellum but are usually darker in colour and quite distinct. When cell walls become discoloured they may be confused with mycelium of *U. nuda,* but this can be checked by examination at x50 or higher magnification.

Common Methods

Reporting results

The result of a seed health test should indicate the scientific name of the pathogen detected and the percentage of infected seeds. The results should be accompanied by information on the test method used, including any pretreatment.

Preparation of Chemicals

Preparation of 5% Sodium Hydroxide Solution

The exact concentration of sodium hydroxide solution is not critical.

Option A

Compound	*g/l*	*g/500 ml*
Sodium Hydroxide pellets	50	25
Cold tap water	1 l	500 ml

Preparation

1. Weigh 50 grams Sodium Hydroxide pellets.
2. Dissolve sodium hydroxide pellets in 1L of cold tap water It is important to ensure that all the pellets

are completely dissolved and this necessitates constant stirring with a metal rod.

Option B: (Sodium Hydroxide can be purchased as 50% stock solution)

Compound	*g/l*	*g/500 ml*
Sodium Hydroxide 50% solution	100 ml	50 ml
Cold tap water	900 ml	450 ml

Preparation

1. Add 100 ml of 50% stock solution to 900 ml Cold tap water.

Lactophenol (1:1:1)

Compound	*g/l*	*g/500 ml*
Glycerine	333.3 ml	166.6 ml
Lactic Acid	333.3 ml	166.6 ml
Phenol	333.3 ml	166.6 ml

Preparation

1. Phenol may be crystalline. To liquefy place in a water-bath and heat gently.
2. Phenol is extremely corrosive, wear rubber gloves and avoid breathing fumes.
3. Add equal parts of glycerin, lactic acid and phenol.
4. Final solution should be clear and almost colorless. The solution will turn yellow with age and expcsure to light. Store in amber bottle and avoid exposure to light.

Safety Precautions

Ensure you are familiar with hazard data and take

appropriate safety precautions, especially during preparation of sodium hydroxide and handling of lactophenol. Dispose of all waste materials in an appropriate way and in accordance with local safety regulations. The extraction and preparation procedure involves the use of potentially dangerous chemicals. Analysts should familiarise them selves with the appropriate material safety data sheets. Appropriate protective clothing should always be worn during the extraction process and at all times when handling sodium hydroxide and lactophenol. Lactophenol should only be handled within the fume cupboard. When preparing the aqueous sodium hydroxide solution, it is essential that the operation is carried out in a well-ventilated room; the analyst should wear full protective clothing.

Treated Seed

This procedure may involve the handling of chemically treated seed. Analysts should familiarise themselves with the risks at tributed to chemical treatments by reference to the material safety data sheets. Contact by inhalation, skin absorption or ingestion must be avoided.

Septoria nodorum

Septoria nodorum is used to the detection on *Triticum aestivum* (Wheat)

Materials

Reference Material - The use of reference cultures or other appropriate material is recommended when ever possible.

Media - Malt Agar or Potato Dextrose Agar containing 100 ppm streptomycin sulphate.

Sodium hypochlorite solution	-	(1% available chorine) for seed disinfection.
Petri dishes	-	When sowing density is given by a number of seeds per petri dish, a diameter of 90 mm is assumed.
Incubator	-	Capable of operating in the range 20°C ± 2°C.

Method

1. *Pretreatment*

   10 minutes in 1% (available chlorine) sodium hypochlorite.

2. *Agar Method*

   Malt agar or Potato Dextrose Agar containing 100 ppm streptomycin sulphate.

3. *Incubation*

   7 days at 20° C in darkness.

4. *Examination.*

   After seven days examine each seed by naked eye for slow growing circular colonies of dense white or cream mycelium that often covers infected seeds. The reverse of the colony is yellow/brown becoming darker with age.

Common Methods

Checking tolerances

Tolerances provide a means of assessing whether or not the variation in result within or between tests is sufficiently wide as to raise doubts about the accuracy of the results.

Reporting results

The result of a seed health test should indicate the scientific name of the pathogen detected and the percentage of infected seeds. When reported on an ISTA Certificate results are entered under other determinations. The results should be accompanied by information on the test method used, including any pretreatment.

Preparation of Media and Solutions

1. Sodium Hypochlorite Solution

Sodium hypochlorite for pretreatment of seed can be prepared from commercial bleach diluted to 1% available chlorine. The concentration of chlorine in commercial bleach varies considerably. Use the formula $V_{stock} = V_{final} \times C_{final} / C_{stock}$ (where V= volume and C= % available chlorine) to calculate the volume of commercial bleach stock solution required to prepare sodium hypochlorite solutions for use in seed pretreatment. To prepare a 1 liter solution of sodium hypochlorite containing 1% chlorine from a stock of commercial bleach containing 12% available chlorine:

$$V_{stock} = V_{final} \times C_{final} / C_{stock} \quad V_{stock} = 1L \times 1/12 = 0.083$$

Thus add 83 ml of the 12% stock to 917 ml water.

2. Malt Agar

Compound	*g/l*	*g/500 ml*
Malt Agar	45	22.5
Deionised/Distilled Water	1 l	500 ml
Streptomycin sulfate	1 mg	0.5 mg

CCP Malt agar constituents should be equivalent to those of the following manufacturers Difco, USA or Oxoid, UK.

Preparation

1. Weigh out ingredients into a suitable autoclavable container.
2. Add 1000 (or 500) ml of deionised/distilled water.
3. Dissolve powdered Malt Agar in deionised/distilled water by stirring.
4. Autoclave at 15 PSI and 121°C for 15 min.
5. Allow agar to cool to approx. 50°C.
6. Pour 15-22 ml of molten agar into 90 mm Petri plates and allow to solidify before use.

   Plates may be stored at room temperature in a dry cabinet or at 4°C.

Potato Dextrose Agar

Compound	*g/l*	*g/500 ml*
Potato Dextrose Agar	39	19.5
Deionised/Distilled Water	1 l	500ml
Streptomycin sulfate	1 mg	0.5 mg

Preparation

1. Weigh out ingredients into a suitable autoclavable container.
2. Add 1000 (or 500) ml of deionised/distilled water.
3. Dissolve powdered PDA in deionised/distilled water by stirring.
4. Autoclave at 15 PSI and 121°C for 15 min.
5. Allow agar to cool to approx. 50°C.
6. Pour 15-22 ml of molten agar into 90 cm Petri plates and allow to solidify before use.

Plates may be stored at room temperature in a dry cabinet or at 4°C.

Safety Precautions

Ensure you are familiar with hazard data and take appropriate safety precautions, especially during preparation of media, autoclaving and weighing out of ingredients. Dispose of all waste materials in an appropriate way (e.g. autoclave, disinfect) and in accordance with local safety regulations.

Treated Seed

This method may be influenced by treatment applied to the seed lot. Seed health tests on chemically treated seeds will generally deliver unreliable test results caused by masking or inhibition of the growth of the target organism.

Bibliography

Bakker, J.P., Bos, A.F., Hoogdveld, J., & Muller, H.J., "The role of the seed bank in restoration management of semi-natural grasslands", In: Ravera, O. (Ed.): *Terrestrial and aquatic ecosystems. Perturbation and recovery*. Ellis Horwood series in environmental management, science and technology. Ellis Horwood, New York, 1991.

Black, M. & Bewley, J. D., *Seed Technology and its Biological Basis*, Sheffield Academic Press, 2000.

Bonner, F.T., "Seed testing", In: Schopmeyer, C.S., ed., *Seeds of Woody Plants in the United States*, Agriculture Handbook No. 450. For. Service. USDA, Washington, 1974.

_______________., "Glossary of seed germination terms for tree seed workers", Gen. Tech. Rep. SO-49. New Orleans, LA. USDA Forest Service, Southern Forest Experiment Station, 1984.

_______________., "Storage of seeds: potential and limitations for germplasm conservation", *Forest Ecology and Management*, 1990.

Peterson, J., "Seed-testing Procedures for Native Plants", In: P.J. Langkamp ed, *Germination of Australian Native Plant Seed*, Inkata Press, Melbourne, 1987.

Poggenpoel, P.V., "Collection and pretreatment of seed from black wattle (*Acacia mearnsii*) seed orchards, Wattle Research Institute, University of Pietermaritzburg, South Africa, Report for 1977–1978. 85–90, 1978.

Schmidt, L., *Guide to Handling of Tropical and Subtropical Forest Seed*, Danida Forest Seed Centre, Denmark, 2000.

Scholer, E. and Stubsgaard, F., *Seed testing*, Lecture Note No. C-8. Danida Forest Seed Centre, Humlebaek, Denmark, 1989.

Simak, M., "Testing of forest tree and shrub seeds by X-radiography", *In: Tree and shrub seed handbook,* Chapter 14, International Seed Testing Association (ISTA), Zurich, Switzerland, 1991.

Stein, W.I., Slabaugh, P.E. and Plumber, A.P., "Harvesting, processing and storage of fruits and seeds", *In: Seeds of Woody Plants in the United States,* Agriculture Handbook No. 450. For. Service, USDA, Washington, D.C., 1974.

Williams, E.R., Gunn, B., Reynolds, D. and Westcott, M., "Germination tests for small collections of Acacia seed", *Seed Science & Technology,* 1992.

Index